TRIBAL CROP-LIVESTOCK SYSTEMS IN SOUTH-EAST INDIA

Tribal Crop-Livestock Systems in South-East India

DAS KORNEL

MANOHAR
2006

First published 2006

ISBN 81-7304-668-9

Published by
Ajay Kumar Jain *for*
Manohar Publishers & Distributors
4753/23 Ansari Road, Daryaganj
New Delhi-110002

Typeset by
AJ Software Publishing Co. Pvt. Ltd.
New Delhi-110005

Printed at
Lordson Publishers Pvt. Ltd.
Delhi-110007

Distributed in South Asia by
FOUNDATION BOOKS
4381/4 Ansari Road, Daryaganj
New Delhi 110002
and its branches at Mumbai, Hyderabad,
Bangalore, Chennai, Kolkata

Dedicated to

DR. GIRIDHAR GAMANG

Contents

Foreword

This book on Tribal Crop-Livestock Systems in South-East India is a timely one. Dr. Das Kornel has done a superb job of assembling a vast amount of valuable information on tribal agriculture in general and crop livestock integrated production systems in particular. In many parts of India, crop livestock integration has been both a way of life and a means to livelihood. There has been a symbiotic relationships between human families and ruminating animals for thousands of years. This has led to different forms of worship of farm animals, particularly cows and bullocks.

Dr. Das has described in this book the rich tribal knowledge of sustainable farming practices and the use of Ethno-veterinary medicine. Unfortunately, such knowledge is now becoming dying wisdom. Such dying wisdom has also led to vanishing crops and animals. It is necessary to preserve such knowledge and genetic resources and to use them wisely through ecotechnological approaches which are based on blending traditional wisdom and frontier science.

I am confident that this book will stimulate a ecotechnology movement which will help us to promote the productivity, profitability and sustainability of crop-livestock integrated farming systems in perpetuity. We owe a deep debt of gratitude to Dr. Das Kornel for this labour of love.

UNESCO Cousteau Chair in Ecotechnology & Chairman M.S. Swaminathan Research Foundation Chennai, India

M.S. Swaminathan

Preface

This book is the result of my work with the tribal communites of Koraput. The objective is to provide factual information, to introduce the reader to the approach to tribal farming, to the wisdom, skills, and the knowledge base of the tribals on the methodology of their cropping and farming systems, and the role of women in the agro-economic-biodiversity setting. It outlines their transition from being a forest-based economy to farming. The description is based on the structural and functional approach to farming as a system required for a quantitative study. There are certain repetitions, which could not be avoided.

The focus here is on the different farming systems followed by the different tribes of Koraput, under different ecological and topographical situations, and their major food crops.

All the examples discussed in the book have been taken from areas I have visited and where I was in a position to confirm the information cited in the relevant literature.

The undivided Koraput district has 56 per cent of population from 51 different tribal communities who are under scheduled tribe category. The individuality of almost every community is reflected in its farming practices. They have evolved and maintained 1750 and more rice germplasm varieties alone in Jeypore tract of Koraput. Such is the rich heritage they have. This agro-biodiversity has been recorded to the extent possible.

For the complex task of writing this book, I have been greately assisted by several tribal farmer friends, who willingly discussed their work and problems. They, in fact, have provided the information that is presented in this book. I have been most influenced by Prof. M.S. Swaminathan's works, for years. I was following his ideas, which have been expressed in his publications from time to time. During 1970s and the 1980s, I used to regularly tune into the weekly Agricultural Round-up programme on the Voice of America. This radio programme had a profound influence on my works, which were designed for local relevance and contained a world vision for agriculture, including bio-diversity and development.

The most valuable advice came from Helge Brunse, Chief Advisor, Danida Livestock Projects through hours of discussion about development. I am also grateful to Peter Lund for giving me the insight into the traditional animal husbandry from his experience and our month long work in Bastar.

I am most indebted to Chresten Petersen, Hon'ble Minister Counsellor, Royal Danish Embassy, New Delhi, for giving me all possible support, opportunity and encouragement to complete this work.

My thanks are also due to Jens Rasmussen, former Secretary, Royal Danish Embassy, who had given me all encouragement to work on traditional agriculture.

I acknowledge with grateful thanks to R.C. Choudhury, IAS, former Joint Secretary, Government of India, Ministry of Agriculture, Department of AH & Dairying, New Delhi and Sanjeeb Chandra Hota, IAS, Agriculture Production Commissioner, Government of Orissa, for giving me the opportunity to complete this work.

I am grateful to Danida's Kursuscenter, Hornbaekhus and to MSSRF Chennai who helped with related literature used in the present study.

I am grateful to B. Lakshminarayana Rao, Ramakant Patnaik, A. Laxmana Rao and Sukant Pradhan for typing the draft.

Jeypore
May 2005

DAS KORNEL

India
Districts of Orissa
BIHAR
WEST BENGAL
N
Sundargarh
Jharsuguda
Mayurbhanj
Deogarh
Kendujhargarh
Baleshwar
Bargarh
Sambalpur
Nuaparha
Sonapur
Angul
Dhenkanal
Bhadrak
Jajpur
Bauda
Balangir
MADHYA PRADESH
Cuttak
Kendrapara
Nayagarh
Khordha
Jagatsinghpur
Phulbani
Puri
BHUBANESHWAR
Kalahandi
Ganjam
Naba-
rangapur
Rayagada
Gaja-
pati
Koraput
Malkangiri
ANDHRA PRADESH

Introduction

The loss of food security promises to become the defining focus of the threat to the global environment in the coming years (Brown,1995). Hunger and malnutrition, which is closely associated with poverty, is still widespread in the world. The increase in population has given rise to concerns about the capacity of the earth to grow more food to meet the challenge. The focus here is on areas and people, mostly tribals, who are rich in the indigenous knowledge systems (IKS) and whether they have been successful.

Indigenous Knowledge Systems

Indigenous knowledge is defined by Warren and Cashman (1989) as the sum of experience and knowledge of a given ethnic group, which forms the basis for decision-making in the face of familiar and unfamiliar problems.

Describing the complexity of traditional agriculture production systems, Carroll et al. (1990) have outlined the sophistication of the knowledge of the people who manage them.

IKS are visualized as a dynamic conception of culture (Warren et al. 1995). They have emphasized the culture as the foundation on which decision-making takes place.

Traditional Farming Systems

Many researchers have stressed the need for sustainable development, especially in food production in the Third World. Alteri and Anderson (1986) advocated a strategy for small farm development in the Third World, emphasizing preservation of traditional farming systems while maintaining biological and genetic diversity at the same time. They further stressed that basing agricultural development on indigenous knowledge, technology, and social organization can provide important guidelines for the design of cropping systems that allow low-income farmers to produce subsistence and cash crops with minimum dependence on external inputs.

According to Caroll et al. (loc. cit.), both social and biological scientists have argued that traditional farming exhibits important elements of sustainability. It adapts well to its particular environment, relies on local resources, it is small in scale and decentralized, and tends to conserve the natural resource base. In practical terms, it is the sustainable agricultural models that combine elements of both traditional and modern scientific knowledge that need to be developed.

However, Warren, et al. (loc. cit.) maintain that, of late, the cultural dimension of development has become a concept in the international development strategy. According to them, new ways are being sought to incorporate cultural components effectively into development plans and programmes. In this context, Kornel (1999a) has documented the cultural dimension of livestock in various festivals and ceremonies of life, covering 10 major tribal groups of Koraput district in Orissa.

Tribal Agricultural Development Strategies

To date, the agricultural development strategies have mainly been based on Western technological solutions, with mixed success rate in developing countries (Den-Biggelaar, 1991). He opined that the Farming System Research (FSR) was advanced as a way to increase the use of indigenous knowledge of farming to make new technologies more adaptable and appropriate to farming conditions. FSR has helped focus attention on people and their knowledge by increasing their participation in problem identification and new technology validation.

Warren et al. (1989) of the opinion that IKS offers a relatively low-cost were approach with potentially high benefits.

Schafer (1989) has suggested ways to utilize indigenous knowledge, along with development of cooperative research and extension programmes in developing countries. Based on the experiences of the World Bank and IBRD in promoting growth, alleviating poverty and protecting the environment, Vanek (1989) said that the new found wisdom of indigenous knowledge is to reconcile their economic and environmental ceremonies.

Warren and Cashman (1989) proposed that IKS can play a key role in the design of sustainable agricultural systems, increasing the likelihood that the rural population will accept, develop and maintain innovations and interventions. They have suggested the following six development scenarios:

1. Overcoming biases in extension communications,
2. Improving existing production systems,
3. Incorporating new technology into the systems,
4. The design activities of research and development programmes,
5. Change from within societies.
6. Creation of an indigenous technological foundation

IKS also suggests close cooperation between the social and agricultural scientists. The women in IKS need to be given an appropriate status. From the experiences of sub-Sahara African social system development, Cashman (1991) concluded that women continue to be excluded from the thinking process. Thus muted women result in loss of women specific IKS and their utility.

In relation to the indigenous knowledge of farmers, Gupta and Pal (1997) studied various models of technology generation transfer in agriculture. They concluded that models that are eco-friendly, culturally compatible and low-cost in nature are most suitable.

Owing to the growing population of the world, food production has come under increased pressure, leading to the gradual disappearance of numerous IKS related to natural resource management (Rajasekharan et al., 1990). Therefore, knowledge-led strategy rather than a material-led one, which attempts to build upon people's own creativity and resourcefulness has been advocated (Pastkia and Hart, 1997). This strategy seeks to preserve the biodiversity and indigenous knowledge related to its use.

The Tribal Farming System, Biodiversity and Green Revolution

Most tribal farmers of our time are rural people, living in the developing countries. In this context Swaminathan (1996) pointed out, 'It is now widely acknowledged that tribal and rural women and men have not only conserved biodiversity of great economic, cultural and social value, but have also enhanced them through selection and value addition through knowledge and information'. He further wrote, 'It is thus appropriate to conserve not only agro biodiversity, but also traditions which led to the enrichment of our knowledge on available genetic material.' Vavilov (1951) advocated the concept of centres of diversity and initiated explorations and germplasm collection programmes. He also helped to add information on the centres of diversity of domesticated plants.

In India, the 'Green Revolution' was triggered through high yielding varieties of wheat, rice, maize, sorghum and pearl millet. In the 1980s, however, Swaminathan continued to visualize and impress upon the necessity to increase food production through biodiversity. He also advocated serious efforts to conserve these most valuable assets for the human population.

The population in tribal belts, which owns the genetic diversity, is also facing serious concerns due to population growth and shortage of cultivable lands. In some pockets, poverty is quite apparent. The governments of these regions are bound to lift them socially and economically. Once these communities adopt high yielding varieties, the traditional genetically diverse but low yielding varieties will be replaced. This was also predicted by Vavilov.

However, the improved varieties of the Green Revolution were not adopted by tribals, though their neighbouring farmers had cultivated these for more than 25 years. This non-adoption was related to social and religious concepts and did not go well with their farming system and food cycle. The tribals living around Bangladesh refugee settlements since 1960, in Umerkote zone of Koraput district, have adopted the improved agriculture techniques only from 1996 onwards.

The Madras Declaration (1996) emphasized the need to move away from a 'favourable area' and 'favoured crops' mindset, to enlarge the contents of the food basket by including under-utilized plants and neglected ecosystems.

This calls for augmenting the resources for conservation, and sustainable and equivalent use of agro-biodiversity. This will reduce the sufferings of poverty stricken population of the country.

The UN Conference on Social Development in Copenhagen, in March 1995, recognized the fact that, today, there is a universal consensus on the need for incorporating the principles of gender and social equity in all spheres of economic development. Inequitable development will not be sustainable in the long term. It was further emphasized at the conference that sustained and equitable development will lead to social integration.

Warren and Cashman (loc. cit.) have proposed designing of a sustainable agricultural system without biodiversity management.

M.S. Swaminathan's model however, advocates sustainable agricultural development along with traditional wisdom and includes agro-biodiversity. This is a challenge. The wisdom and knowledge in tribal agriculture needs to be incorporated, along with *in situ* conservation of biodiversity, through participatory planning and gender equity in a given society and area.

Food and Tribal Farming Systems and the Future

There are certain areas in the hill tribal farming system where periods of food scarcity occur. They may last for about 2 months of the year, from late summer to pre-monsoon period. The period is overcome by use of non-conventional food resources. Moreover, during this period the families actually need good food to live up to the demands of hard physical labour in the fields. This also is the period when livestock do not get much feed from grazing. Present day, limited period of food scarcity loss in tribal life may be attributed to the missing link of forest resources.

The small millets are the life support crops. These crops are the 'hope factor' on which they gradually build through the food scarcity period, and finally return to the happy days of the food cycle, when there is plenty at home.

There are also famine saviour plants in the same system. The food scarcity period is supported by wild growing fruits, mangoes and jackfruit, and edible bamboo shoots. In some places kendu fruits, forest mushrooms and other wild roots, berries and leafy vegetables also come under this category.

The small millet group of crops are produced in the shortest possible period. They support the most difficult food shortage and scarcity periods in the annual food cycle, especially for the primitive and also hill dwelling tribal groups. These species of plants also tend to escape drought. These valuable genetic resources need proper attention, as seen from tribal point of view. The tribals have selected early maturing crops of legumes and cereals also. This agricultural biodiversity has been developed and conserved by tribal families on the basis of their ecology and farming conditions. These have been respectfully maintained through generations. The religious concept binds the family to maintain these supporting traditions and heritage.

The livestock and poultry that they rear at home, have short maturity periods, short gestation and breed prolifically. They are most suitable in such situations. The livestock and poultry in tribal regions are very special. They can reproduce fast, grow early and provide the family with food and cash.

These assured food supply systems developed by tribals, each with a different personality of its own and suited to its own local conditions, need to be respected. The development plans built under such system *in situ*, will be the answer to those societies who are in transition from traditional farming to improve farming system. The successful systems of food production can well be transferred to regions and societies that require it to survive. These small but most essential production systems will be most useful to alleviate poverty.

Koraput District—A Profile

The Koraput district is the southernmost end of Orissa. It has between 17° 48' N and 20° 7' N latitude and between 81° 24' E and 84° 2' E longitude.

The largest district in Orissa till 1992, Koraput was divided into four districts—Koraput, Nabarangpur, Rayagada and Malkangiri in 1992. Here in this book, Koraput district denotes the undivided Koraput district.

The total area of Koraput district is 26,961 sq. km., with 42 blocks. The population density is 111.5 persons per sq. km., against the state average of 236. The tribals comprise about 55.86 per cent of the total population (Census 2001). The district population has continued to grow at about 20 per cent between 1981-91 and 1991-2001.

The population of Koraput district is 35.87 lakh (Census 2001), against the Orissa state total population of 367.07 lakh.

The details of population in different administrative districts and undivided Koraput district are given in Table 1.

TABLE 1: ADMINISTRATIVE DISTRICT OF KORAPUT AND POPULATION DETAILS, 2001

District	Total population in lakhs	Scheduled Tribe in %	Scheduled Caste in %	Density per sq. km
Koraput	12.22	55.53	13.39	12.3
Malkangiri	5.07	66.50	19.90	69.0
Nabarangpur	10.08	55.26	15.11	160.0
Rayagada	8.50	54.99	14.28	94.0
Undivided Koraput	35.87	55.86	15.67	111.5

Forest Cover

Koraput has a forest area of 8,87,000 ha (as per 1994-5 ESR, GoO). The total forest area in the state is 57,18,000 ha.

Podu Land (slash and burn cultivation)

Orissa has a total podu cultivation coverage of 1,84,018 ha, of which Koraput district has 30,651.16 ha of podu at present. It is far below the area that was estimated to be under forest cover in 1958 by Money, the Conservator of Forests, Orissa. His estimate was 12,80,000 ha.

Farm Holdings

The average size of the holding is 1.88 ha. The average size of operational holding of major farmer groups of Koraput is:

Marginal Farmers	0.59 ha
Small Farmers	0.59 ha
Semi-Medium Farmers	2.80 ha
Medium Farmers	6.01 ha
Large Farmers	14.41 ha

Land Tenure in Tribal Areas

The community owns part of the land, which is divided among the villagers for cultivation. The rest is owned individually. Very few of these land holdings are legally settled. Women have no legal rights over land. The Land Alienation Act, which applies to the tribals, restricts them from selling land to the non-tribals, without approval of the government.

Land Classification

Land classification of Orissa state is comprehensively classified by Lenka, 2001. The land classification forwarded by him for Koraput district hill region is Dongar I, II and III, followed by Jhola.

Dongar is the land situated on the hill slopes, with Dongar I being the uppermost. It is stony, steep and used as podu cultivation land. Located below Dongar I, the Dongar II land is used to grow the niger, upland rice, suana and arhar crops. Below Dongar II lies Dongar III, where maize, upland rice, niger, and pulses, including *arhar* are taken up. The fertility here is better than Dongar I.

Jhola is the land from where flood water flows during rains. The farmers, carefully prepare the land with suitable bunds and divert the excessive and perennial water, as required.

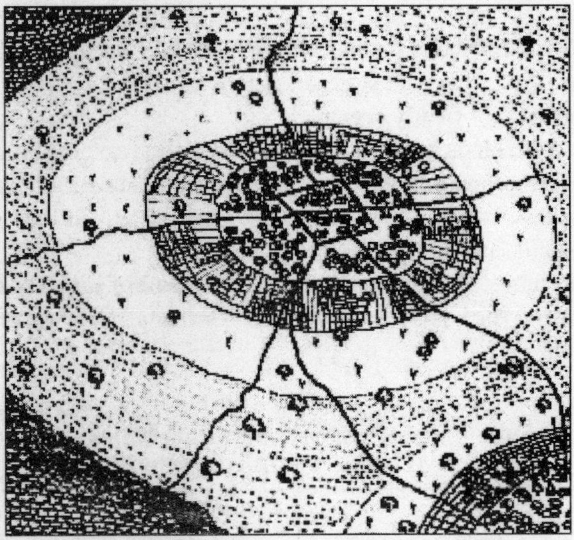

Fig. 1: Spatial organization of land use in blocks plain land of Koraput (1) House and garden (2) Permanent cultivation (3) Intensive fallow system (4) Bush and forests

Fig. 2: Spatial organization of land use in hill blocks of Koraput (1) House (2) Byre (3) Manure pits (4) Garden (5) Jhola paddy (6) Podu cultivation and (7) Upland permanent cultivation

The land in the plains is called 'Beda', and has three classifications. The Beda III is the lowest stratum, where most water stands still and which is useful for paddy crops.

Attal are the lands on the banks of rivers Kolab and Indravati. They are fertile. A good number of vegetable crops and sugar canes are grown in these lands. The Bhattras are very skilled in low moisture crops of winter and summer in attal lands.

Koraput has 5.37, 2.67 and 1.24 lakh ha land of high (I), medium (II) and low types (III) respectively. The total cultivated land is 9.28 lakh ha.

TABLE 2: ADMINISTRATIVE DISTRICT-WISE KORAPUT CULTIVATED AREA ('000 HA) DURING 1995-6

	High Land	Medium Land	Low Land	Total	% high land to total
Koraput	166	78	58	302	55
Malkangiri	78	40	23	141	55
Nabarangpur	144	42	29	215	67
Rayagada	138	43	22	203	68
Total	526	203	132	861	–
Percentage	61	24	15	–	–

Rainfall

The annual rainfall in Koraput district in cm is available since 1901. It was having an average of 148.65 cm (1901-96) with variance of 25.72 and c.v. in percentage is 17.30.

The decennial rainfall variation of Koraput was reported by Lenka, loc. cit. In Koraput the maximum was also in the decennium 1951-60 (165 cm) and decreased to 141 cm (14.5 per cent) during the decennium 1971-80.

Koraput district had medium droughts in 1901, 1920, 1923, 1935, 1941, 1960, 1964, 1968, 1969, 1974, 1976, 1979, 1987, 1989, 1993 and 1996(16). There was severe drought in 1965 only.

The Koraput district had medium floods during 1910, 1914, 1917, 1919, 1925, 1936, 1953, 1956, 1959, 1971, 1977, 1990(12), whereas it had severe floods during 1970 and 2001 only.

Variation of Rainfall in Koraput District

The pre-monsoon sets in by the first week of June, and normal monsoon sets in by 15 June. The monsoon causes heavy precipitation in the hills and the western slope of the hills. As the current passes over the hills, the intensity of rain decreases. The areas lying east of the hills are called rain shadow areas and receive less rain. The Rayagada receives 152-250 mm rains, which is much less than Koraput. In monsoon months, the rains continue for days together. This was the usual course in Jeypore 20 years back.

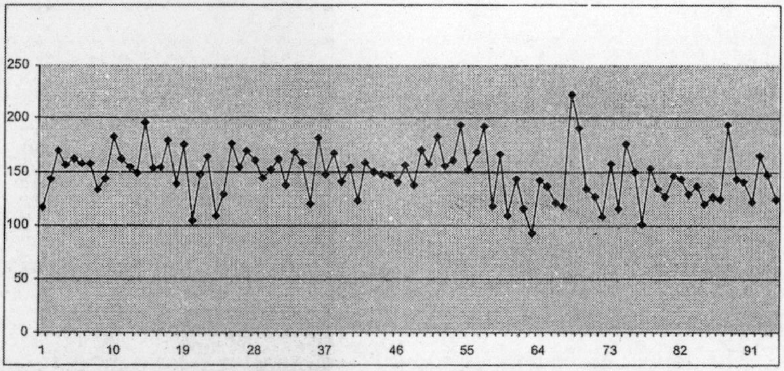

Fig. 3: Annual Rainfall in Koraput district, Orissa (1901-96)

The present rainfall data analysis shows that the rainfall in Koraput district has decreased after 1975, to the order of 15-20 cm in a year. Before 1970, the rainfall in July and August in Jeypore, Nowrangpur, Malkangiri and Koraput region varied between 425 and 500 mm. By 1966, it has been reduced to 320-400 mm level. This has serious consequences. These results co-relate to reserve forest land reclamation to rehabilitate Bangladesh refugees in Malkangiri and Umerkote regions. Since, then serious proportion of timber harvest is continuing both officially and non-officially. Still after 1980's serious efforts have been made by soil conservation department, forest and socio-forestry to plant more green plants.

It was during Bangladesh refugees rehabilitation in Malkangiri and Umerkote regions of Koraput district as on 1978, 1,08,004 ha and additional area of 42,652 ha thick reserve forest was turned to cultivable land from 1979 to 1983. This is single and largest forest coverage cut down to rehabilitate refugees in one district.

Ambient Temperature of Koraput

The mean annual temperature of Koraput is the lowest 22.2^0 C in Orissa.

TABLE 3 : THE MEAN MAXIMUM AND MEAN MINIMUM TEMPERATURE OF KORAPUT DISTRICT (^{0} C) FROM 1901-97

Months	max T	min T	Mean
January	28.6	05.8	17.2
February	31.9	08.3	20.1
March	34.8	13.0	23.9
April	36.7	17.4	27.1
May	37.6	17.6	27.6
June	34.9	18.1	26.5
July	29.2	17.2	23.2
August	28.5	18.2	23.2
September	29.1	17.6	23.4
October	29.3	13.3	21.3
November	28.0	06.6	17.3
December	27.2	05.3	16.3
Annual mean	31.3	13.2	22.3

The annual mean temperature of Jeypore Rice Research Station is 3^0 C more than that of Koraput which is due to height variation. The Koraput is 912.2 m high above sea level.

Plate 1: Hill jhola land

Plate 2: Wet land rice farm, Kotpad

The Tribal Life: Forest, Food at Home

The basic need of the tribals is food and clothing. An essential prerequisite for the planning of the tribal area is a close understanding of the relationship of the tribals with their environment. It has been observed that apart from a few groups that are better off, most of the tribals depend on the forest for fruits, flowers, roots, tubers, etc., for a part of the year. This is because the agricultural produce and minor forest produce, wherever collected, do not provide the families with subsistence throughout the year. The most crucial aspect in tribal development plan strategy is to pull them out of the brink of starvation and provide food security throughout the year. This can be achieved through higher returns from existing agricultural produce, forest products and higher productivity, as well as higher production in the agriculture, horticulture, animal husbandry, forestry and small industries sectors. As mentioned before, during the transition period, tribals rear small animals, subsisting on the food and cash the occupation brings. To evade further exploitation by local moneylenders, tribals must be provided with financial security. This can be achieved through self-help groups or village commitees.

Forest in Tribal Life

The forest area in Koraput district is rapidly decreasing. The current vegetation represents the bio-edaphic series and very little of the primarily forest is left. Now there are only the mixed deciduous forests of Shorea-Tectona-Diospyrox. The vegetation on the ground consists of a large number of species, many of which provide food to the tribals. As mentioned earlier, Grigson (1949) has observed that, 'Famine has never been a problem in Bastar as half of their (tribal) food is delivered from forests.' He further wrote that, 'The difficulty is not so much to say what they collect and eat as what they do not; in fact, if one were to take one of the lists of trees, shrubs herbs compiled by Central

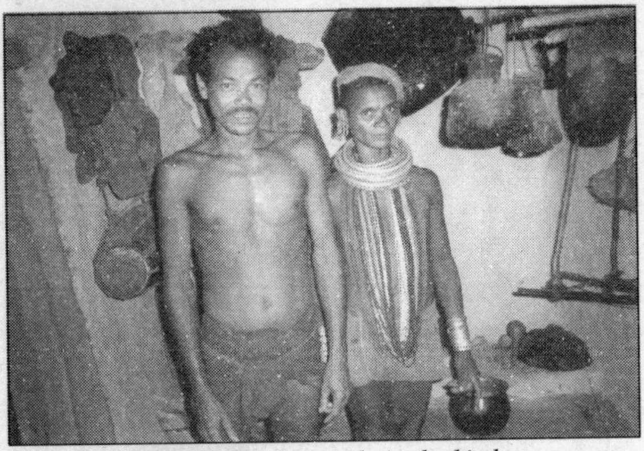

Plate 3: Bondo couple in the kitchen

Plate 4: Bondo women, distilling mahua liquor

Plate 5: Sago palm tree with pot

Plate 6: Landa preparing herbal ingradient, Padua

Province Forest products, one would not even then produce an exhaustive list.' Elwin (1947), Roy and Rao (1957), Jain (1963), Tiwari (1977), Randhawa (1980), and Fernades et al. (1988) listed forest species of plants that were as food by tribals. Ethnobotanically, however, the knowledge of food habits of various tribes is still fragmentary and incomplete.

Observations suggest that a tribal does not cut a tree, particularly a fruit tree, though he may be obliged to fell some other tree in certain compelling circumstances.

The tribals derive employment, food, minor forest produce etc. from the forest. Above all, they obtain fuel from the forest for cooking, heating and even lighting purposes. Thus, the tribals are still in the integrated, non-specialized stage of economy, which depends on the forests for food and shelter. The forest lands are rapidly losing out not only to agricultural land but also to some species of fast growing trees that are used for industrial purposes. Somehow, *Eupatorium odoratum* is appearing, as the major weed covering the land and forest of Koraput district. The lantana on high altitude is the major weed problem.

For long, the tribal economy has subsisted on minor forest produce (MFP). The ASCSI, Hyderabad had organized a study about the role of MFP in Madhya Pradesh, Andhra Pradesh and Orissa states. It was concluded that on an average, income from MFP accounted for 10 to 60 per cent of the total of tribal families.

The tribals are now in the development stage, essentially a problem of transition. Sharma (1985) cautioned that the conceptual distortion could arise from the fact that forest-based economy, agricultural economy and urban economy represent three different stages in human development and each one has little application to each other.

The Tribals' Food and Drinks

Chiefly among traditional tribal societies, there is an intimate connection between the food, natural resources and the farming system. The later two elements influence the status of family food cycle and supply.

Usually, tribal families have three regular meals a day at home. However, this routine is altered during podu cultivation and other field operation days. On such occasions, the whole family, including children, eat their lunch in the field itself. All the cooking is still done at home.

At about 10 in the morning they eat boiled rice and few cups of mandia pej (ragi). Pej, their staple food, is made of rice, ragi or small millets, with ragi being the predominant ingredient. Ragi pej has a special method of preparation. The clean ragi grain is ground into flour and soaked in water for two days. After that, rice is put to boil in a pot. When it is half cooked, the ragi flour is added and the pot is filled with water. Then, rice and ragi are cooked together. This preparation is called pej. At about 2 in the afternoon, they have a few cups of stale pej. At 8 in the evening, they again have boiled rice, if available, or ragi pej. Pulses usually form a part of the evening meal.

Plate 7: Red ants Chutney

Every meal includes chutney, made out of tamarind, mango or mahua. Meat and fish may be added to roots or vegetables, which are cooked with chilli and turmeric. These delicacies vary from day-to-day, but the main diet remain unchanged.

Some Delicacies

Young people go into the forest in groups, to search for nests of red ants. These are usually located high up in the trees. When they find such a nest, one of the young men climbs up the tree and breaks off the branch on which the nest is made. One of them plunges his hand into the nest and quickly fills the leaf cups, with ants and their eggs. They brush off the eggs and closely tie up the ants in the cups. These little bundles are carried back to the village, and put over fire. After roasting for few minutes, they open the cups and smash the ants into a paste. Some salt and chilli is added to the ant paste and the mixture is placed back into the fire to bake. This delicious chutney is generally had with liquor or food.

They also spend hours digging out field rats and mice. They roast rats over the fire, to remove the fur before eating. The Bondos, who like rat meat, fatten rats for the purpose. The tribals in general are very fond of dry fish, specially salted dry sea fish. The riverain dry fish is also available. They purchase dry fish from weekly markets. It is roasted over fire and consumed as chutney. Some tribes, on hills as well as in Bastar, roast frogs as a delicacy. The non-poisonous snakes are also eagerly sought, although the demand is now decreasing. Some people believe that snake meat has medicinal value. For instance, consuming python meat is said to be a cure for a pot belly.

Cooking of Food

While women usually cook at home, it is common for men to cook at feasts. The rice is cleaned by women, who also make leaf cups and plates for the feast. The men cook in pots placed over fire, which is lit in a long trench. When rice is ready, it is tipped out in a great pile on newly made bamboo mats. For the feast, a pig is killed and roasted on the fire till all the hair is burned off. In some tribes the pieces are made with skin, in some it is deskinned. The skin, meat and bones are cut into small squares of equal size. It is important that this is equally shared. These pieces are fried in *mahua* oil, if available, or any other edible oil and then put into the pot, which also contains rice water, turmeric, salt and chilly. It is boiled till ready. In some tribes, the animal meat is simply boiled in water with salt and turmeric, if available. The fish is also boiled or fried in oil as above. In some tribes, fowls are cooked in leaf cups. In most cases, the intestine of the chick is squeezed out, and the whole fowl, with head and feet, is roasted with little salt. The eggs are also baked in the fire, with leaf wrapped around it.

At home, the family sits around the pots of pej, which is taken out straight off the fire and placed in the middle on the floor. There is no such rule that women should eat after the other family members.

Edible Oil in Food

The oil component in tribal food is the least. They use lilnseed (*alsi*) oil, tolo oil and kirchi oil, and some Kondhs of Dasmanthpur area claim use of castor oil in food. Nowadays, some tribals purchase edible oils from the market.

The Kirchi plant grows in local forests. It bears yellow berries, and when decoated, a blood red pulp is seen. It is compressed to extract red, sweet edible oil. The oil is also mixed with *alsi* oil.

Intoxicants and Other Drinks

Among tribals, drinking liquor is both a duty and a pleasure (Elwin, 1947). Drinking the *mahua* spirit is a duty because it is offered to the gods at every festival, and during naming of a child, at marriages and funerals. But the juice of the sago-palm, and even landa rice-beer, is only for pleasure (Elwin loc. cit.).

Sago-palm (*Carupta urins*): This is the common and most popular intoxicant in the tribal world of central India and Koraput. The tree is locally called Salpha. The tree throws flowering peduncles, which resemble elephant trunks and which have their ends cut-off before they flower. It is from here that the juice of the tree drips. It is collected in pots or bamboo tubes, which are fastened tightly or supported by special devices. The pots are emptied every morning and evening. The fresh juice is not intoxicating but is said to be laxative. A little of the old juice is always left in the pot to enhance fermentation. In order to make the taste strong, the Muria tribes add little of the root of the Danbaher tree. The Bondos add barks of various plants, like Achki, Artun, Gosangte, Pitamari, Zukuski, Tangri, etc. (Patnaik et al.,1984).

Small groups of men, go out to the tree, taking with them roasted crabs, little bits of chicken, various kinds of chutney, especially the roasted wild red ant chutney, and enjoy drinks of sago-palm.

Toddy palm (*Borassus flabellifer*) is also called Tad or Tal tree. It is abundant in Malkangiri. Its sap is cooked by cutting the ends of the flowering spadices, like that of sago palm. The sap of toddy palm is also used and processed like sago palms.

Date palm (*Phonex sylvestris*), otherwise known as *sindi* or *khajuri*, is tapped in Gunupur region. An incision is made in the bark, below the crown of the tree, to tap the juice. Its processing and utilization is as above.

Mahua (*Bassia latifolia*) is the common tree in most tribal areas of central India, like Malkangiri, Raighar, and the Gunupur region of Koraput. The dried ripe flowers are high in sugar content, and are distilled for liquor. Some tribals, like Maria, Koya and Bondo, make a wine by distilling and straining the *mahua* flowers several times.

Landa, the favourite intoxicant of Poraja and the tribes from the plains of Koraput, is the cause of half the homicides committed in Bastar (Grigson, 1938). It is made up of equal portions of husked rice and ragi. The rice is soaked in water, strained and then pounded to a paste. The ragi grain is first damped and allowed to sprout, then dried in the sun for three to four days. It is further soaked, strained and pounded like the rice. Both are then well mixed. Then, an earthen pot, which has holes in the bottom, is covered with a little of bamboo slices, on which siadi leaves are placed. The mixed rice and sprouted ragi are piled over these leaves. The pot is then placed on top of another pot, which contains water. The latter is placed over the fire, so that the contents of the upper pot can be well steamed. After having been steamed long enough, they are mixed in another pot with water to the consistency of pej. It is then placed in a cool corner of the house, to ferment for five or six days. The mouth of pot is covered with leaves. When it starts emitting a foul smell, it is an indication that the landa is ready.

The landa is prepared in large quantities to serve guests in ceremonies like marriages, functions and other festivities.

For certain tribes, like the Hill Maria, Peranga and others however, it is taboo, possibly to avoid food scarcity.

Pendom, a beer, is another popular intoxicant made from rice or minor millets. The rice is cooked and spread over a bamboo mat to cool. Many add fermenting and flavouring agents to enhance the process. These are sold in weekly markets as square tablets of one inch, or like chapati cakes. Many do not disclose the names of the plants used for the purpose. The mixture of cooked rice is collected in a steamed pot, and a little water is added to it. Then, it goes into fermentation process for three to four days. Thereafter, the pendom is strained before being offered to guests. It is served in leaf cups to guests on festive occasions and ceremonies. It is a very popular intoxicant in summer.

The Bondos make two types of rice beer, 'li', the best drink of all and very delicious and intoxicating, and pendom, which corresponds more or less to the Bastar landa (Elwin, 1950).

The best 'li' is made from old rice that has been roasted three months previously, and is placed in a special jar to mature. The women collect the barks of *Bassia latifolis, Terminalia tomentosa* and *Diospyros melanoxylon,* and the roots of *Phoenix sylovestris* and *Baliospermum axillare*. These are cut into small pieces, which are then put into an earthen pot full of water. A second earthen pot, which has a small hole at the bottom, is closed with a bamboo sieve, in which the rice is placed. The first pot with water is placed over the hearth, with the second pot over it. Thus, the process of distillation starts with the rising steam, and the 'li' starts to collect.

The pendom differs from 'li' in being some what simpler to make and in requiring an artificial fermenting agent, instead of the various roots and barks employed in 'li'.

Among most primitive tribal groups, like Bondo, Koya, Dongria Kondh, and other hill tribes, liquor is prepared from various ripe fruits. It is prepared from ripe tamarind, banana, jackfruit, mango juice, *khajuri kuli*, *mahua* fruits, and figs, etc.

Food Preservation

Different tribes follow different degrees of food preservation. On the whole, the art can be concluded as primitive. Every tribe keeps some reserves of grains, animal products and vegetables whenever possible. *Mahua* flower is dried and preserved for future food by Koyas. The Poraja and Kondh, Koya, and Bondo preserve animal meat. Some tribes even collect from dead animals specially during epidemics in poultry. These are cut into small pieces, undried, smoked and kept in airtight bamboo containers. Sometimes, I have seen dried animal meat, which is most unhygienic and unfit for consumption, being preserved. This is due to poverty and scarcity of food. The Pengos, Parenga, Bondo and Porajas take special care to dry vegetables in season, either in the sun or at the hearth, and carefully store them for future use of the family.

History of Agriculture in Koraput

There is no consolidated report about the history of agriculture in Koraput. This depends on the different tribes that came here for various known and unknown reasons. Where they came from, and when, is not well established and is largely based on speculation and the folk tales of the tribals themselves. Since 150 BC, many dynasties ruled the area, whose history is recorded from Kharavela. From time to time the river bed and land attracted many invaders and occupants from various regions.

There are ponds in plain villages. The hill dwellers depend upon the perennial springs for drinking water and limited irrigation.

The remains and findings of Puspagiri at Podagadh of Umerkote, which are well established by historians to be of Nala dynasty and their once capital, is dated to the sixth century AD. The ruins of the capital, which is closely linked to the Vainavism cult, have ponds.

Jains, Ponds, Wild Rice and Fruit Trees

The ponds, specially as twins in the plains of Borigumma onwards, are most probably a part of the religious culture adhered to by Jains in this district, since the fifth and sixth centuries AD. They grew wild paddy in one of the ponds, which still exist in the Kotpad and Borigumma regions. Small boats were used to harvest the paddy. Such processed wild rice, called *Titing chuda*, is still sold in festivals. In the other pond, they grew lotus flowers, which were also offered to the deity as offerings. Village communities now use these ponds for pisciculture.

The Jain monks brought to Koraput the religious culture of planting fruit bearing trees all around. It was the duty of Jain students to plant these trees during monsoon, so that travellers, hungry people and villagers could enjoy the fruits. In the tribal society, this concept is still alive. In the plains of Jeypore tract and in the Gunupur region, fruit trees were mostly the result of cultural influence of the Kalingas. Gunupur also had some Buddhist influence.

Twin Tanks

Tanks were first excavated in the Tamil country in early seventh century AD, by the Pallava kings and then Mahendravarman I. During their glorious period in the tenth century AD, the Cholas also excavated tanks. The chain tanks system was followed in Karnataka during the same period. A similar system is found in the B. Singhpur, Borigumma and Nowrangpur tracts. This area was ruled by the Cholas and the Chalukyan rulers during the ninth century AD. These chain tanks had scientific layouts, as in watershed management systems.

Taboos to Plant Trees

Some sections of the society were forbidden from planting certain crops and fruit trees, which had economic significance. Loss of clan, misfortune and untimely death were some of the implied consequences of defying these taboos. These sanctions were imposed by the privileged upper classes, who were unwilling to share the economic benefits accruing from these crops and trees.

This is well established in one district in north Orissa, where a priest is required to conduct a religious ceremony before anyone can plant coconut, arcanut and other fruit trees. The bamboo plantation and ownership was denied to all but for one class of the society. Although the practice still exist, it is on the deline.

In some parts of Koraput, it was a strong belief among people that planting coconut, mango and jackfruit trees could bring the family misfortune and loss of clan, etc. However priestly class could plant. This of course was a part of keeping the working class people needy; and administration so as to prevail upon them, few privileged class of the then rulers had control over the issue.

Similar taboos are attached to a perennial creeper (*Coccinia indica*) locally known as the *tundri creepers*. In the plains, all tribal households plant it in the backyard. The water that is used to wash hands and utensils is not allowed to go waste and is further used to water the vine. The plant bears enough fruit to provide the family with a regular vegetable supply, at least on alternate days. Daily consumption of the vegetable, however, is said to lead to foolishness and an ill-developed brain. But this plant is of great economic importance and is a persistent supplier of food to every family. This is very similar to *Sajana* or *Muringa* plant in the coastal Orissa. During the rainy season, it often gets infested with pests.

On hills, most people depended on naturally appearing fruit trees, while in the plains, people relied on the technique of planting new trees. The system, however, is now changing.

The sago palm trees were possibly first planted by tribals to enjoy its sweet intoxicating cool sap during hot seasons. Everyone takes care of the sago palm sapling, wherever they are planted.

Malis, who came from Kosala were probably the first pioneers of vegetable cultivators in Koraput. They are skilled in intensive irrigation base cultivation of vegetables, and have in place a systematic sharing process of spring water and land.

The ragi crop, which is the staple food of most tribes, is locally well known as *mandia* probably came to this land from Karnataka and Tamil country through the invading armies of the Cholas and perhaps also the Western Chalukyas. Perhaps the same is also true of the *alsi* oil crop (*Guizotia abssinica*) which is also seen in neighbouring district of Bastar. It is presumed that the ragi is named after Mandya region in present-day Karnataka, the place from where it originally came.

Origin of Rice and Koraput

Sharma (1996) wrote, 'As far back as 1936, when Dr. K. Ramaiah, the rice breeder of Madras Presidency, visited the Jeypore tract, he was amazed at the genetic diversity of the area and suggested that it might be the centre of rice origin.' It was thus in 1947 that a Rice Research Station came into existence at Jeypore, which is now with Orissa University of Agriculture & Technology, Bhubaneswar.

The ricebeans (*Vigna umbellata*), a pulse crop, is locally known as Dongar rani. It means the queen of hillocks. It is believed that Koraput district is the secondary origin for rice bean (Mishra, 2000 personal communication).

Introduction of New Spices and Plant Varieties

The pineapple crop of Dongria Kondhs in Rayagada area was said to be the contribution of zamindars of Jeypore, who had brought it from

Burma. Radhaballava Samantray first introduced cabages during 1927 in Arku Valley, while Sardar Sunder Singh introduced vegetable crops like *methi*, carrot, turnip and sugar beet to Jeypore in the 1930s. During 1964, improved rice varieties like Taichung and Tainan 3 were introduced in the district. First chemical fertilizer application through extension appeared in 1955, while pesticides were introduced during 1970s. Sir Vikram Dev, had introduced the concept of horticulture, as well as village mango orchards in the Jeypore zamindari.

Pulla Reddy, ICS, Dewan of Jeypore State established two large size fruit tree (tropical) orchards, one at Mahulbhatta near Jeypore and another at Mahaliaput, the south end of Jeypore, which had almond or *badam* trees (*Prunus amygdalus*) from Kabul, coconuts, mangoes and others. This was a large collection. He also established one coconut garden in Jeypore, called Nadia Bada. These stand damaged today.

The cashew plants were first introduced by the Jeypore zamindars and were prevalent around the Jeypore Hastishala (elephant sheds in 1950s). The credit for large scale plantation of cashew through soil conservation programme, around Jholaput area of Koraput district, goes to the late Shyamananda Patnaik. Later this proved to be the best induction of economic plant in all of Orissa.

The sandalwood (*chandana*) tree is special in Koraput. Its plantation was taken up between Jeypore and Koraput on the hills and at Nandapur by the zamindars during their regime. The cane though limited is only specific to Ramagiri forest of Koraput.

The Mestha (inferior type of jute) crops were introduced in Malkangiri and Umerkote regions by Dandakaranya project during the 1960s for the Bangladesh displaced persons. The commercial production of crops such as maize, mestha and *til* were the contributions of the Dandakaranya project to the district.

The cultivation of Virginia tobacco (*Nicotina tobbaco*), organized by a company in 1964 in Dumriput, in Koraput, did not go well. Owing to the drought in 1948, the government of Orissa started two agricultural farms, one at Pottangi and the other at Dabugaon. In 1967-8, the Pottangi farm was later transferred to Orissa Agriculture University and Technology, for the cultivation and test of cardamum and cloves. Late in the 1970s the coffee board initiated coffee farming in the region. Although the production was good, it was hampered by inadequate marketing. But it is being taken up again, on a limited scale, by farmers from Kerala, who have recently settled here. The first coffee plantation was taken up at Mahulbhatta by Jeypore zamindars.

The monsoon potato cultivation was started in red loomy soil during 1966 and 1967, and commercial cotton cultivation was started in Rayagada during 1974.

The Avocado trees (Butter fruit) were introduced by a missionary group during 1995, followed by Danida-Integrated Livestock Development Project (ILDP), Jeypore, which introduced it in Malkangiri region during 1995. A small number of bread fruit trees were also introduced in the region.

The Semiliguda Mixed Farm successfully introduced the peach fruit trees. In the 1980s, they had introduced the cherry fruit plants into the area from Kerala.

The Central Cattle Breeding Farm, Sunabeda introduced oats into the area in 1975, and fodder crops like NB21, Australian soobabul trees, para grass, and Coimbatore 1-hybrid napier during 1990. Danida-ILDP introduced Bracharya group of grasses during 1993.

The Koraput district along with western Orissa form the boundary between the south Indian short tail sheep and north Indian long tail sheep.

Wild Fowls, Wild Buffaloes, Elephants and Livestock

The district has good population of wild fowls in jungles; the one in north of Koraput and Rayagada onwards are *Gallus sonneratii* (Grey jungle fowl) but around Koraput and down south are dwarf *Gallus gallus* (Red jungle fowl). The poultry populations in the Kondh tribal villages on hills are small in size and contain genes of local wild fowls. It is opined that probably one of the early domestication of fowls was also from Koraput and Kalahandi districts. There are numerous folk stories among tribals about the origin of fowls, including that of artificial incubation of wild fowl eggs (Elwin, 1954).

Koraput is also a source of Arni wild buffalo (*Bubalus bubalisarni*). There are instances of wild buffaloes cross-breeding with domestic buffaloes, with the Manda being an example of one such cross-breed. From the socio-religious association of all the Austro-Asiatic tribe groups of Koraput, who sacrifice buffaloes in most of their festivals and ceremonies, it appears that it was probably they who had introduced domesticated buffaloes in Koraput (Kornel, 1999b).

This part of Koraput was also a great source for fine war elephants. It was one of the prime reasons for many dynasties in the past centuries to acquire this land.

Cattle are mostly small in size and in descript type. Most cattle population in Malkangiri area have small humps or less marked hump which is attributed to the past migrating herds of Baiparis or Lambardies of north India to south India.

Manda buffalo breed is native to Koraput and found around Kakarigumma and Panchamali hills. Paralakhemundi breed of buffaloe breeding tract extends into Gunupur. Motu breed of cattle, milch strain herds are found between Kalimela and Motu of Malkangiri area. Another moderate milch population of cattle found around Gunupur area is yet to be recognized. The Goudos of Koraput have few homogeneous cattle herds, with black colour splashed over white bodies. These are part of the small population of inbred cattle, maintained by individual families for generations.

The bullock cart was first introduced in the Koraput district in 1865 (Fewcash, 1907).

Introduction of New Breeds of Livestock and Poultry

A small flock of white leg horns was first introduced at Borigumma during the 1950s, by the government of Orissa, followed by brown leg horns at the Koraput Poultry Farm. Red Sindhi and Harayana breed bulls were introduced during the same period. The first Jersey bulls were to Orissa through the Hill Cattle Development Scheme, Koraput, during 1966. In Orissa, the first artificial insemination of cows, through frozen semen, was carried out in 1975. In 1992, a calf was born through an embryo harvesting programme at the Central Cattle Breeding Farm, Sunabeda. The first full-fledged embryo transfer and embryo freezing laboratories of eastern India were established during 1993 at the same place.

Agro-Biodiversity and Koraput

The biodiversity resources represent a most precious asset in natural resources available to human kind. The maintenance of both biological and genetic diversity is fundamental for the proper functioning of any ecosystem and they are expected to make important contributions to social and economic developments (Khoshoo, 1987).

Vavilov (1951) has given the basis of variety and diversity in plants and agriculture. He opines that many of these secondary centres of genetic diversity are located in tribal belts and has expressed concern that the governments of these regions are bound to lift them socially and economically. Once these communities take to high yielding varieties, the traditional and genetical diverse, but low yielding varieties will get progressively replaced. These varieties are repositories of many useful genes brought together by thousands of years of selection. This holds true of livestock germplasm also. Swaminathan (1982) has also expressed a similar view about the wealth of genetic diversity in these tribal areas that are now in the wake of scientific, technical and industrial development.

The existence of more than 51 different tribal groups in Koraput district itself, their isolated habitats spread over wide and varied ecological pockets, the practice of *meriah* (human sacrifice) until late eighteenth century, the self-limiting social habits, the natural barriers like dreaded black water fever (Malaria), difficult terrains and hills without proper pathways, the menace of tigers and leopards, dreaded poisonous snakes and so on discouraged the entry of the outside world in this area. These are probably also the main contributing factors behind the origin and conservation of the vast agro-biodiversity existing today (Kornel, 1999b).

Owing to the inroads made by plant varieties and animal breeds better known for their economic productivity, the agro-biodiversity of Koraput district is under threat. Swaminathan (1982) further cautions saying, 'while soil erosion attracts attention since it is visible to the eye, the erosion of valuable genes in animals and plants tends to get ignored until a disease epidemic or some other production problem, from which a genetic solution is necessary, crop up'.

Sharma et al. (1998) of M.S. Swaminathan Research Foundation (MSSRF), Chennai studied and reported the agro-biodiversity of Jeypore tract, Koraput, Orissa. In this limited study it has been recorded that there are 340 varieties of paddy, 8 species of minor millets, 9 species of pulses, 5 species of oil seeds, 3 species of fibrous plants and 7 species of vegetables. MSSRF is continuing this study further.

Jeypore Tract, Centre of Rice Origin and Tribals

Rice has had a diffuse origin (Harlan, 1956), both in space and time. Several centres of origin have been proposed for Asian rice; of which southern India was proposed by Watt (1892), and Jeypore tract in Orissa a secondary centre was proposed (Ramiah and Ghose, 1951; Nair, 1973). Other centres abroad have also been identified. Among centres of origin proposed in India are the Malabar Coast in Kerala, the Jeypore tract in Orissa, and north-eastern foothills of the Himalayas. In these areas there are several wild taxa, including both annual and perennial rafi pogan, and high variety diversity. Sharma (1996) wrote that, the Asian strains of rice originated from India. According to him there are three basic varieties of rice, which are basically responsible for the rest. One is *Ouse*. *Ouse* and *aman* are two separate ecotypes, according to Dr. Moringa of Japan. *Ouse* and *japonica* rice are from the Himalayan region, south China and the South-Eastern Asian hill areas. South-Eastern Asian hill rice is responsible for the origin of *japonica*. *Aman* has originated from *ouse*.

In India, there are three major centres of genetic diversity. Sharma maintains that, 'as far back as 1936, when Dr. K. Ramiah, the rice breeder of Madras Presidency, visited the Jeypore tract, he was amazed at the genetic diversity of the area and suggested that it might be the centre of rice origin. When Vavilov published his book in 1951, he said that north-western Himalayas was the primary origin of cultivated rice. However, Dr. Ramiah called it the secondary origin, and proposed a Jeypore botanical survey. This was funded by ICR (1955 to 1960). Some 1,750 collections were made during that period but the breeding system was not recognized because the *japonica indica* hybridization project was at its peak.

The primary centre of origin is expected to contain many dominant genes, some of which are resistant to disease, insect pests, and abiotic factors. Thus, these land races are very important.

Further, regarding the land races and new breeds, Sharma et al. (1998) wrote that, 'in the early 1950s, the late Dr. K. Ramiah (the renowned rice geneticist of India) proposed that the Jeypore tract could be the centre of origin'.

They further cited Dr. H.I. Oka (1962) who studied the cultivated and wild varieties of rice in this area and propounded the view that the wild rice types of this area (Jeypore tract) have the potential to differentiate into *japonica* and *indica* and, hence, the traditional rice varieties of this area could be called incipient *japonicas* and *indicas*.

Wild rice, *granulata* is prevalent and identified from Ranigada, B. Singhpur, Govindapalli and Durgi areas of Koraput, whereas varieties belonging to *glaberima* was identified in Papadahandi collection of rice. They were named *Jaipureansis* (Govindaswamy and Krishnamurthy, 1958; Govindaswamy et al. 1966).

Out of 1,025 types selected by Govindaswamy and Krishnamurthy loc. cit, 20 were found to have *japonica* characteristics. They were Singapuralu of Kalimela (Malkangiri), Nunidhan of Kotpad, Naikani of Gumma, Biribachela of Gotiguda, Biribasera of Modkiniguda, Bandadhan of Bariabahal, Melidhan of Khumarimundi, Makad of Gotiguda, Madada of Dadkabhata, Badapati of Pedakonda (Malkangiri), Sosaberi of Chiddipali (Malkangiri), Koshabarlu and Sindipuralu of Podia (Malkangiri), Badamangi of Tunbarala, Nanibudi of Banamaliput in Nandapur, Mirlidhan of Kakerigumma, Sansundari of Chuidri, Paikani of Laxmipur and Galurbande of Padamgiri (Malkangiri). These were compared to Norin and Juiho *japonicas*.

The botanical survey of rice by ICR, as mentioned earlier, showed 1,760 varieties of cultivated rice and 150 varities of wild rice. The studies from Jeypore tract indicated *perennis* and *fatua* wild rice types.

This itself can explain the space and time that rice has travelled in Koraput district, along with Austro-Asiatic tribes like Sabara, Bondo, Gutob Gadabas and other groups like Bhattras and Bhumias. They brought the terrace cultivation, of rice with a transplanting technique. Their skill in weed control through water management, is a significant contribution.

The tribal geneticists show amazing knowledge in ragi variability. They have short, medium and long duration small millets, including ragi. They even have varieties named after number of fingers the crop has. These are homogeneous trait in the said variety.

The crop duration, especially the small millets have most bearing on family food needs. It is most essential component of hill tribes to meet the immediate family food demands which appear soon after the summer food scarcity.

As the custodians of their respective families through generations, women have amazingly maintained the tradition of propagating a variety of plants for centuries. The selection pressure continued for high yield and maintenance of pure lines. Tribal women have thus contributed immensely to the conservation of biodiversity in Koraput, till today

Genetic Dents

Several incidents of genetic denting of local livestock have taken place in the past. Grigson (1938), the former Bastar State Commissioner wrote,

> In 1932 a rumour went through all the tribes of Jeypore and spread thence into Bastar that a god had descended on one of the mountains of the Eastern Ghats and commanded all men to give up keeping black poultry and goats, wearing clothes or using umbrellas or blankets with any black in them and using beads or articles made up of aluminium alloy. The message was rapidly bruited abroad and everywhere villagers had 'bohorani' ceremonies for purifying themselves from disease and cast out black goats, cocks, hens, umbrellas, blankets, 'waskats', beads, aluminium ornaments and domestic utensils on the village boundary.

Again, during 1971 in Bastar district and Malkangiri subdivision of Koraput district, a great amount of damage happened to sheep, goat, pig and poultry germaplasm due to the spread of a similar rumour that the tribesmen should renounce non-vegetarian food habits or else the world would perish (Kornel, 1999b). This caused considerable dent to germplasm resources.

But still, the traditional, abiding nature of Koraput tribes has succeeded in the conserving the crop germplasm to a greater extent. One of the examples in this field is the conservation of paddy varieties through a traditional ceremony among the Gutob Gadaba families (Kornel, 1999c). The procedure signifies the seriousness of the ceremony for rice conservation in Gutob Gadaba tribe.

The Koraput and Kalahandi districts have been identified for plant germplasm collection by National Bureau of Plant Genetic Resources of ICAR.

Tribal Groups, Different Crop and Varieties

RICE

The Gutob Gadabas came out with the following few varieties of rice, ragi and other small millets. The rice or *keram are,* Bhatta keram, Kalamara, Kodieser, Modo keram, Sapur keram, Chatialozoi, Mali keram, Angur keram, Patsadi keram, Bayagunda, Laserbutki, Balu keram, Kulur keram, Amagada, Pakanagada, Goriachipti, Kaliachipti, Kendumundi, Chilpa keram, Bukti keram, Bodo kukti, Alsidhan and Kardi keram, etc.

The few rice varieties named by Sabaras of Gunupur are: Ragadhan, Jaipadhan, Katanodadhan, Kanduladhan, Ambadhan and Sanagadhdhan, etc.

The Bhattras of Kotpad have rich rice diversity and a few of the varieties are: Sunakadika, Haldichudi, Longembari, Donaprasad, Dubraj, Parang Kaktichuti, Badsabhog, Ladakhuda, Bagadichudi, Lokatimachi, Kharakoili, Lakatikeyi, Machakanta, Haldisari, Mansa, Pakhiya, Dongar, Badamakuda, Bayagunda, Matia, Assamchudi, Baringi, Lalu (scented), Umariachudi, Mer, Kalamali and Bakatichudi, etc.

Bondo tribes have the long duration fine quality rice germplasm, some of them being: Piriman, Sunalaser, Dabla laser, Laser, Nini, Chiti, Chudi, Daulacer and Tharmang, etc.

The Dongria Kondhs are not rice cultivators in the true sense. They have Jaijati, Mahisar and Sarana Mahisar, which are the long duration type. The Batkuli is the upland type rice and a scented variety called Basanadhan.

The Matia and Koyas of Malkangiri have some special, deep water rice varieties also. These are: Paticha, Bayagunda, Gadakuta, Ishabati, Kalamati, Ratanchudi and Assamchudi.

The Poti, Balinga, Badapato and Singhapuria are also deep water rice varieties. The Mahipal, Bak, Palt, Pandedhan, Satakadhan, Bababuta, Dandkadhan, Kalsidhan, Sapauriadhan are some of the shortest upland rice, which yield in 60 to 70 days.

RAGI (MANDIA)

The Koya in Malkangiri has good varieties of ragi in this region. The Dasarajhara are harvested in September, the Samarkhan variety has four fingers, the Kuruba has five fingers and is the late maturing type. The Narmathanghara is medium duration ragi.

The Dongria Kondhs have two types of ragi, which again based on duration. They are Doramandia which is harvested early during Dussehra (September-October), the other one is called Sitamandia and is harvested late. They classify ragi by the shape of its fingers. The straight type is called *langramandia* and the curly type is called *tusa* or *mudamandia*.

The Gutob Gadabas on the hills have some excellent ragi varieties. These are: Mami Samel, Karanga Samel, Chodigodi Samel, Leda Samel, Poda Keranga Samel, Pitti baska Samel, Machgar Samel, Mundi Gadrada Samel, Marda Samel, Dudkaranga Samel, Kaugodi Samel and Bhatti Samel, etc.

ARHAR (*KANDULA*)

The *arhar*, a pulse crop, is known as *kandula* locally. The Laxmipur-Rayagada-Gunupur belt is a rich resource of *kandula*. There are three types in it. One is large with bold seeds, called Bodo *kandula*. There is another very small seed type (San *kandula*), and is mostly found in Gunupur region.

The other type is called Majhia or medium type. They have red and black coloured seed, as well as mixed coloured seeds, with black, red, and white. The medium and small seed type *arhar* is used as *dal*. The *arhar* plants have good fuel value for the tribals in this region. The plants grow up to 6 and 8 ft in height.

The Dongria Kondhs report four varieties of *kandula*. They are all of long duration and are Chachikanga, Machakanga, Serkokanga and Lelukanga. They are all mixed coloured seeds. The Machakanga grows tall and has a thick stem.

SUANA

The Gutob Gadaba have some of the following varieties of Suana (Iring). Mami iring, Dhabli gurji iring, Kalia gurji iring, Gaja iring,

Modo iring, Kasam top iring, Chittri mundi iring, Jhipal iring, Kasam Dhabli iring, Mati gurji iring, Loda mami iring and Dasara mami iring.

The list is exhaustive. The magnitude of the diversity existing in Koraput is quite evident from the information available. MSSRF, Chennai, has a similar area of interest, particularly the rice genetic resource on which the work is in progress.

The ICRISAT, Hyderabad, has some germplasm collections of small millets and sorghum of Koraput and preserved in their germplasm bank.

SUSTAINABLE USE OF BIO-DIVERSITY

The 'Green Revolution' was initiated by breeding new, high yield varieties of wheat, rice, maize, sorghum and pearl millets in Asia. The breeding was successful due to the effective use of genetic variability found in nature. Genetic homogeneity in the cultivated as well as the inbred lines of hybrids is gradually increasing. In recent years, adoption of genetic engineering has effectively utilized the genetic variability both in plants and animals. Global concern about conservation and sustainable and equitable utilization of bio-diversity, led to the concept of agro-biodiversity at the Convention on Biological Diversity (CBD), adopted at the UN Conference on Environment and Development (UNCED) at Rio de Janeiro in June 1992.

Conservation of Agro-Biodiversity and Tribal Family Rights

Further, the technical conference convened by the FAO in Leipzig, Germany in June 1996 stressed on the action to be taken for conservation and sustainable utilization of plant genetic resource for food and agriculture. The contribution of the families of indigenous tribal and rural communities to sustainable food security, through their continued efforts in the area of plant and animal genetic resources conservation, should be recognized and rewarded.

The Genes and Future

The modern gene-recombination and gene-transfer technology depend heavily on traditional genetic resources to draw resistant genes for disease and pests, greater tolerance to environmental stress, draught, salinity, water-logging and extreme temperature regimes for plants. However, among livestock, genetic answers to disease resistance, ecto and endo parasite tolerance, gradual body fat reserve depletion during droughts, and seasonal fodder constraints, water-logging condition, escaping ability from predators, fertility independent of nutrition and higher conversion of roughages, etc., will be sought. These genes are available to greater extent in centres of diversity, like the tribal regions. The Koraput district is one such resource rich area.

The Wild Races of Plant Genetics in Koraput

The wild relatives of the following plants are prevalent in Koraput district: wild rice, wild bananas, wild brinjals, wild round gourds (*lau*, which are called *anka* locally), wild okra, wild melons, wild *arhar*, wild *tundri* or little gourd and many more yet to be identified. There may be some more wild species of other edible plants, which require identification.

- Wild rice is still found in the ponds of Kotpad-Borigumma regions, even on hills of Banuaguda.
- The wild bananas are found on the hills, specially from Laxmipur onwards as well as in the Govindapalli region. These have small, round and hard seeds. The pulp is very sweet when ripe. The tender bananas, with immature seed, are cooked.
- The wild brinjal is edible. They are medium to small in size, and mostly grown in the home garden by Kondhs, near Raphkona.
- The wild bottle gourds are still cultivated on a limited scale on the hills of Koraput by all tribes. It is bitter and not edible. It serves as a vessel to carry hot gruel to the field and even drinks. The Bodo Poraja have a special ceremony in their calender to harvest the new wild bottle a gourds.
- There is wild okra in the Bondo and Didayi hills forest. The fruits are small in size and are edible.
- Wild *arhar* (*Atylosia cajanifolia*) is seen in Semiliguda-Pottangi forests, and the same has been collected and maintained by Central Soil Research Institute, Semiliguda.
- The Sabara tribes utilize wild melons as food (Elwin, 1955).
- The wild *tundri* or wild little gourd is used as food in Kotpad region.
- The wild sajana (Muninga) is bitter, and is found in Bondo hills.

The other Plant Germplasm of Future Interest

- Attal brinjals of Kolab and Indravati riverbeds are large round, and green, with white star bottoms. These grow in least moisture conditions. The plant is resistant to many disease conditions. This is a local plant resource. There are several other vegetable and crops specially known as attal tomato, attal sugar canes and other crops. There is another brinjal type found in Nowrangpur, which is cultivated during summer and sold during July. It is also a local germplasm. A small sized brinjal, grown on hills of Padua and extending to the Eastern Ghat, is of future interest to plant breeders from disease resistant point of view.
- The Nodia rice, a unique variety with two grains in single seed found may be of future interest (B.M. Samantray, personal communication).
- The tobaccos of Koraput, is the age-old Virginia type. It is found both in the hills and the plains of the district. The Kondh, Bondo and Koya tribes have good old lines.
- The extra large and heavy pumpkins of Malkangiri. There is a lot variability in shape, size, colour of skin and pulp of the local pumpkins. The pumpkin type of Borigumma-Kotpad needs attention. Pumpkins are planted in the *kharif* season.
- The old papaya line is found in the villages. These have long years of productive life, and bear large fruit. The Kondhs, Sabaras have some good lines of early maturing papaya.
- Bondos have good variability both in shape and size of the local bottle gourds. The Didayi tribe have long bottle gourds, as well as round constricted neck and round bottle flask type which is common in wild bottle gourds.
- Bondos, Kondhs, and Koyas have special plantain varieties both for vegetable and table. The following table varieties of banana are seen in Dongria Kondhs. They are Pattadi, Velutadi, Kiljatadi, Katuwtadi, Kumutitadi, Podhasingi and Muguni. The Pattadi has a dark red cuticle on fruits and Muguni is large in size; however, both fetch high price. Katuwtadi and Kumutitadi are best for vegetable cooking. The Velutadi and Podhasingi are used for ceremonial purposes.
- The Matia tribe in Mathili area, as well as other tribes in Malkangiri grow the ash gourds extensively. It deserves attention.

- The little cluster tomatoes still exist in places, specially in Bondo hills. It is an old line.
- The sweet potatoes and yellow sugar cane of Mali farmers are other local crops of interest.
- The 'bailo' beans grown by Rayagada Kondhs and Dongria Kondhs need attention. These are of two types. The variety grown on hill lands is called *Gutubalam*, and the beans are large in size. The bailo grown on the plains are called *Unubalam*.
- The *rajma* and jhudungo beans of Pottangi are said to be native to Koraput (H.M. Samantray, personal communication). The seed colour shows variability. The size of the seeds, as well as early to late maturing varieties are of interest in jhudungo.
- There is a large diversity among creeper beans called Semi. This is apparent from the colour of the flowers, pod size and shape. The seed colour, shape, and size show variation.
- The smallest chillies which have a strong taste, is called *Dhan Maricha*. These are dark purple in Malkangiri and grown by Koyas. It is perennial. It is green in Dongria Kondhs of Bissamcuttack. This ideally suits the daily need of the family. The chillies of Dongria Kondh are in high demand and sold at high a price.
- The *Karamanga* creeper of Gutob Gadabas is a single source. It is of interest as a vegetable, and equally suitable for a pickle preparation. This is special to the tribe.
- The Bhattras and Dongria Kondhs cultivate *Semili Konda* (Cassava), which they claim to be native to their region.
- The 'Khamba Alu' locally called '*Nangal Konda*', grown in the tribal homes of the plains of Koraput, needs attention. The Bhattras, Dongria Kondhs, Sabaras and Bhumia tribes have good lines of it.
- Dongria Kondhs, Sabaras on hills, Didayi and Kashipur Kondhs are rich in cucumber genotypes. It is an important food source for them. Dongria Kondhs have two types of cucumber. One is called Dhonda mundra, which is long, white and thick as a thigh, while the other one is Kanchkapoda. Roundish, it is white but its head has a touch of yellow.
- As mentioned earlier, the little gourd from the plains of Koraput is a very useful plant for vegetables to a family, and needs to be compared with other improved varieties. It is a predominant crop of Bhumia and Bhattara.

- There are at least three types of castors, the *Chikti Jada*, the *Badha Jada* and *Ranga Jada*. These are very old lines, available throughout the year. The seed is used for extraction of castor oil by local methods.
- The '*alsi*' has several varieties and distributed over varied eco-geography in Koraput district, however they are not found in Nyamgiri hills of Dongria Kondhs. *Alsi* has wide genetic variation, thus selection to improve can be very effective. A good deal of work has been carried out in Regional OUAT Research Station, Semiliguda.
- Cultivating rice and fish together was a well-known farming method in Malkangiri region by the Koyas. The rice varieties thus grown are special and well known elsewhere.
- The Kuttings or rice bean, otherwise locally called 'Dongarrani', is a legume crop of future interest. It grows well on hills under least soil fertility and acidic soil conditions. It is a creeper, with profuse foliage and growth. The yield of pods is good. Usually its seeds are light green in colour, but the Didayi tribe has dark coloured seed variants.
- Small millet germplasm is varied, both in types and varieties. They are ultra short to long duration crop. Ragi is the staple food of tribals in Koraput district and it shows wide genetic type. The millet duration factor is of prime interest in the food production cycle of tribal group and they have retained it since centuries.
- The Sorghum is of interest in the Laxmipur-Nandapur belt. Pearl millet is seen in the Kashipur-Gunupur belt. The Sorghum is still popular in these regions.
- The Suana is common, but good high yielding varieties are seen in the Kashipur-Rayagada belt. There is a great deal of variability in Suana crop. The Dongria Kondhs have four varieties of Kosala. These are Kodrukada, Sindhitode, Burluki and Edingineri koladu. They prefer the last variety, which has bluish seeds. The Burluki is white, while the other two are black after their sēeds mature.
- Both the varieties of the grain amaranthus, red and white, are found on Eastern Ghat hills near Sunki group villages and are conserved by Kondhs.
- There are old lines of 'popcorn' maize varieties and cultivated in the backyard farms. The Didayi tribe has local maize type which mature very early.

- The opportunity in mango, jackfruit and tamarind genetic resource is numerous. A local indigenous tree, the mango exhibiting a great genetic diversity in terms of with different coloured tender leaves, flowers of different shades and colour, different sized fruits and fruit cuticles, pulps of different grades of colour, and unripe fruits with no sour taste. The most remarkable aspect is the longevity of the mango trees.
- The jackfruit is of two types, *Lalia* (slimy) and *Khaja* (non-slimy). The size of the fruit, number of fruits, colour of fruit skin and pulp, varied shapes are the main attributes. Besides there are trees, which bear fruit on main trunk, some on stems and some on both the sites. Aroma and taste are the other different attributes. There are early and late fruit bearing trees. Two centres of genetic-diversity of jackfruit in Koraput is known. They are Tanginiguda, near Boipariguda and Kurli of Bissamcuttack. This author carried out a detailed study to identify the types of jackfruits for the Tanigniguda centre. The study was supported by photographs.
- The tamarind is well distributed in the hills and plains, and so also in the Bastar district. There is variation in the size, shape and weight of the fruit pulp recorded.

The trees live for very long. These three trees, e.g. mango, jackfruit and tamarind of economic interest need to be studied and documented further. The people can themselves select right kind of trees and propagate. In the coming years, local selection of varieties in terms of economic gains, disease resistance and long productive period will appear. Without a doubt, the early fruiting types will be preferred in the future.

The Kondhs have pioneered turmeric, specially the rich pigmented one for centuries. The tuber crop, collected from forests of Koraput, deserves serious attention. The Central Tuber Research Institute - Regional Centre Bhubaneswar has collected species of tuber plants from Koraput and Bastar, which are claimed to be most superior.

The contiguous patches between Gajapati district and Gunupur-Koraput, inhabited by Sabara tribes, have a large, distinct variety of plantain and citrus fruits. This requires detailed study for further conservation.

Biodiversity of Livestock and Poultry

The prolific breeds of sheep, goat, pigs and poultry are most interesting. The goats and sheep here mature far earlier than those found in the neigbouring coastal plains. They have very short breeding intervals. The prolificacy trait in these animals is independent of nutrition, as evident among local pigs. The poultry is also early maturing. These are the genetic needs of a world that is increasingly under pressure to meet food requirements of its growing population.

The livestock and poultry biodiversity of Koraput, Ganjam and Kalahandi districts of Orissa have been clearly drawn up by Kornel, 1988 and 1999b.

Plate 8: Medium cluster branch fruits

Plate 9: Large, trunk fruiting-Kundra

Plate 10: Large cluster

Plate 11: Large, branch fruiting

Plate 12: Fruiting, trunk, Nandahandi

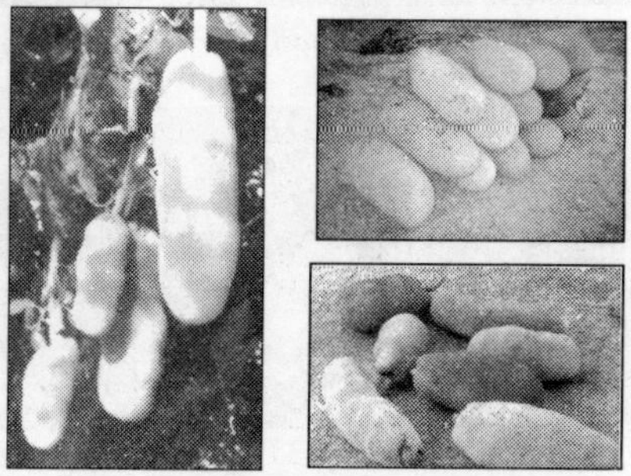

Plate 13: Long and heavy fruit—light green

Plate 14: Branch profuse fruiting

Plate 15: Medium uneven fruit body

Plate 16: Jackfruit market, Kundaie

Plate 17: Extra large fruits, Kotpad

Plate 18: Dongaria Kondh

Plate 19: Didai

Plate 20: De sai Kondh

Plate 21: Bondo

Plate 22: Koya

Plate 23: Easternghat hill brinjal

Plate 24: Attal Brinjal small

Plate 25: Attal Brinjal medium

Plate 26: Attal Brinjal large

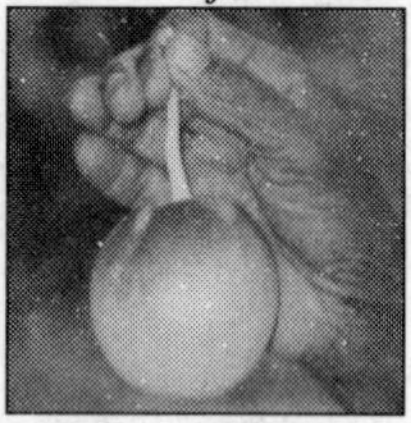
Plate 27: Wild Brinjal edible

Plate 28: Large green spotted, Kakrigumma

Plate 29: Large round, pink, Kunduli

Plate 30: Long pink, Boipariguda

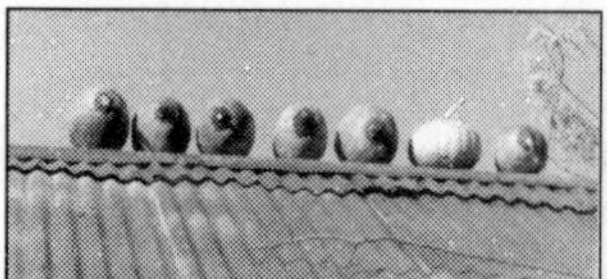
Plate 31: Conservation of punpkin with cowdung seal on the stump

Plate 32: Green, White, Yellow Paukali

Plate 33: Medium pink, Jeypore

Plate 34: Sundarmania bamboo

Plate 35: Edible bamboo

Plate 36: Pataka bamboo, Borigumma

Plate 37: Semi, seeds, multi-colour, Kondamberi

Plate 38: Dongria Kondh Dhana Maricha (chilli)

Plate 39: Karamanga, Gutub Gadaba Bottlegourd diversity

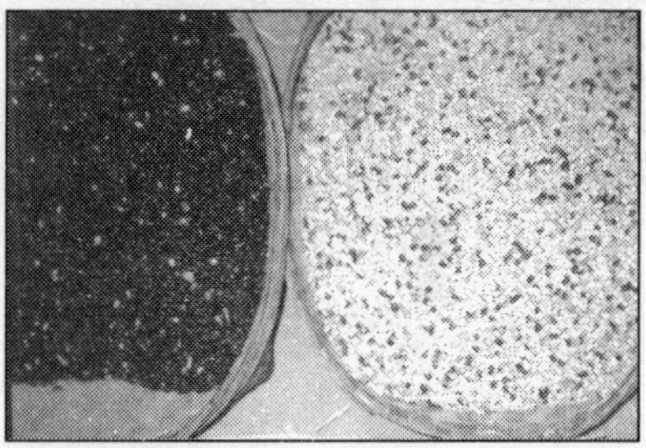

Plate 40: Black and White seed coat horse gram

Plate 41: Kunduli

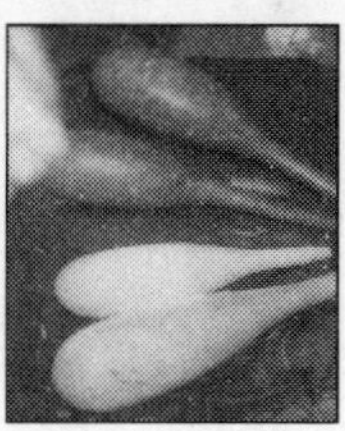

Plate 42: Paukali

Plate 43: Baligaon

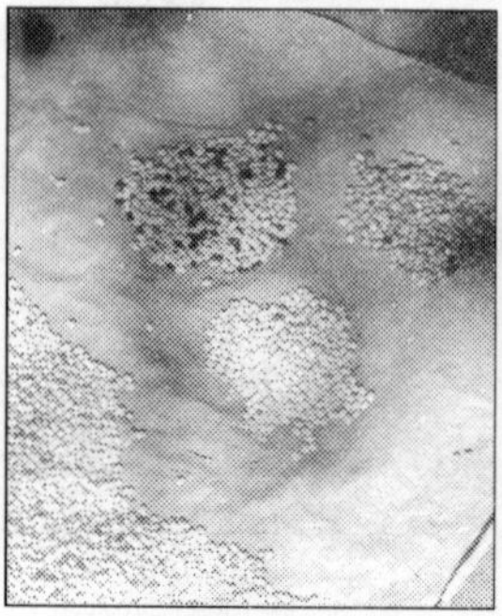

Plate 44: Jhota multi-colour seed, Padua

Plate 45: Jhota multi-colour seed, Kuumulumma

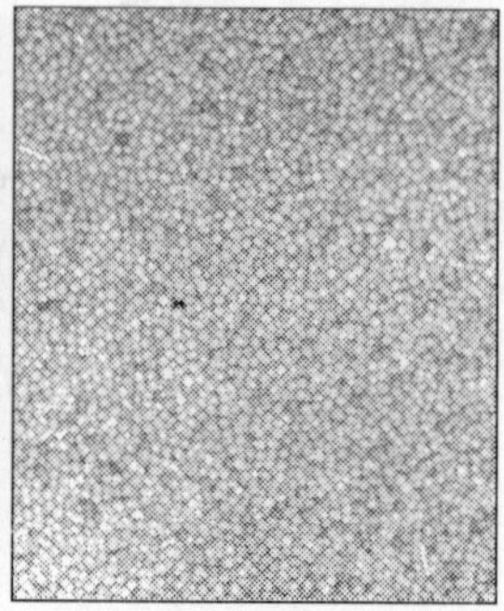

Plate 46: Chona (local-peas), Kashipur

Plate 47: Jhudunga multi-colour seeds, Kudumulgumma

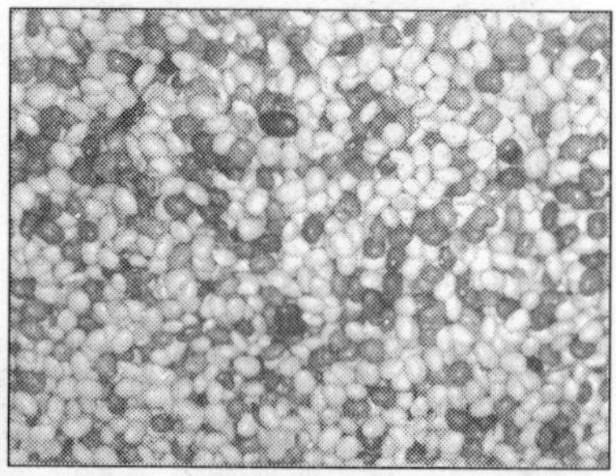

Plate 48: Bailo multi-colour seeds, Rayagada

Plate 49: Kadula multi-coloured seeds, Rayagada

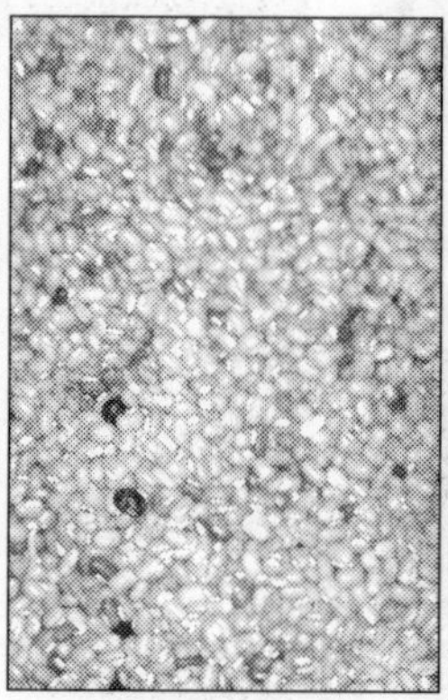

Plate 50: Rice bean seeds Kudumulgumma

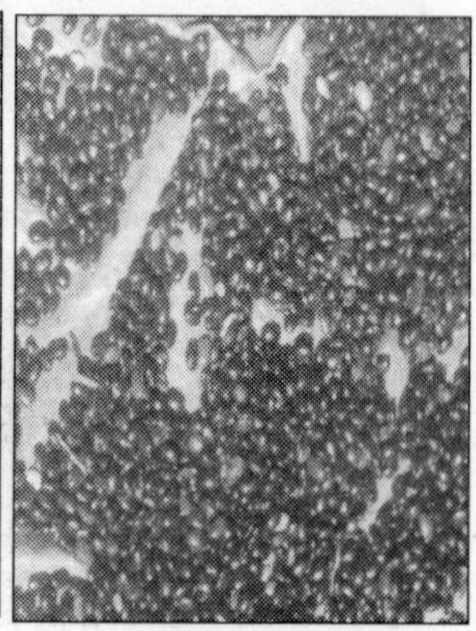

Plate 51: Mali Jhota (Dolichos), Padua

Rangada

Sabola

Ghee Sabola

Biribania

Dhabala

Kandulbania

Biri-kandul bania

Hitkaranyan

Badlia

Chitra

Paunsia

Kalahandi poultry breed, Bisamcuttack

Plate 52: Vezaguda Poultry

Some Tribal Groups of Koraput and Agriculture

It has been recorded that there are 51 scheduled tribes inhabiting Koraput district alone. The Kondh, Sabara, Koya, Gadaba, Bhattra, Bhumia are the major tribe groups of Koraput district. The following distribution does not count very minor tribes.

Koraput District—Block-wise Distribution of Scheduled Tribe Population

KORAPUT

1. Koraput Block	(a) Poraja (b) Kondh
2. Semiliguda Block	(a) Poraja (b) Kondh (c) Gadaba (d) Ollar Gadaba, (e) Parenga
3. Pottangi Block	(a) Kondh (b) Poraja (c) Gadaba (d) Ollar Gadaba (e) Kotia, (f) Parenga
4. Nandapur Block	(a) Poraja (b) Kondh (c) Kotia (d) Gadaba (e) Pengo, (f) Parenga
5. Lamtaput Block	(a) Gadaba (b) Poraja (c) Kondh, (d) Parenga
6. Dasmanthpur Block	(a) Poraja (b) Kondh
7. Laxmipur Block	(a) Kondh (b) Poraja
8. Narayanpatna Block	Kondh
9. Bandhugaon Block	Kondh
10. Jeypore Block	(a) Poraja (b) Bhoi Gadaba (c) Halva/Pentia (d) Bhattra (e) Omnatya (f) Bhumia
11. Boipariguda Block	(a) Bhumia (b) Poraja (c) Dhurva (d) Kondh
12. Kundra Block	(a) Bhumia (b) Poraja (c) Bhattra (d) Kondh (e) Dhurva (f) Pentia/Halva

13. Kotpad Block	(a) Bhattra (b) Dhurva (c) Omnatya
14. Borigumma Block	(a) Poraja (b) Bhattra (c) Omanatya (d) Pentia/Halva

MALKANGIRI

1. Malkangiri Block	(a) Koya (b) Matia (c) Poraja (d) Dhurva (e) Santala (Refugees)
2. Korukonda Block	(a) Koya (b) Matia (c) Bhumia
3. Khairput Block	(a) Bondo (b) Gadaba (c) Kondh (d) Didayi
4. Kalimela Block	(a) Koya (b) Konda dora
5. Kudumulugumma Block	(a) Poraja (b) Kondh
6. Podia Block	Koya
7. Mathili Block	(a) Koya (b) Bondo (c) Poraja (d) Bhumia (e) Dhurva (f) Halva

NABARANGPUR

1. Nowrangpur Block	(a) Bhattra (b) Poraja (c) Pengo
2. Kosagumuda Block	Bhattra
3. Nandahandi Block	(a) Bhattra (b) Poraja (c) Omanatya
4. Tentulikunti Block	(a) Poraja (b) Bhattra (c) Pengo
5. Papadahandi Block	(a) Bhattra (b) Poraja (c) Pengo
6. Dabugaon Block	(a) Bhattra (b) Poraja
7. Umerkote Block	Bhattra
8. Raighar Block	Gond
9. Jharigaon Block	(a) Bhattra (b) Poraja (c) Gond (d) Kondh (e) Bhunjia (f) Halva
10. Chandahandi Block	(a) Gond (b) Bhunjia

RAYAGADA

1. Gunupur Block	(a) Kondh (b) Sabara
2. Gudari Block	(a) Kondh (b) Sabara (c) Jatapu Dora
3. Ramanaguda Block	(a) Sabara (b) Jatapu Dora (c) Kondh
4. Padmapur Block	(a) Sabara (b) Kondh (c) Jatapu Dora
5. Chandrapur Block	Kondh
6. Bissamcuttack Block	(a) Kondh (b) Dongria Kondh

7. Muniguda Block	(a) Kondh (b) Dongria Kondh
8. Rayagada Block	(a) Kondh (b) Jatapu Dora (c) Konda Dora
9. Kashipur Block	(a) Kondh (b) Poraja (c) Pengo
10. Kalyansinghpur Block	(a) Kondh (b) Dongria Kondh
11. Kolnara Block	Kondh

The Austro-Asiatic tribes are Sabara, Gutob Gadaba, and Bondo. The Parengas and Dongria Kondhs can be also classified on the basis of their physical characteristics to some extent. They are the oldest inhabitants of the district. The Austro-Asiatic tribes brought rice cultivation and buffalo culture to the district. The Koyas who are of the Gond lines, have cattle-based economy, and their life and living centered on *mahua* tree base. The Bhattras and Bhumias showed excellence in cultivating rainfed rice and least moisture adoption crops. The great skills of Dongria Kondhs in horticulture-based integrated crop cultivation, steep hill cultivation of the Pengo, and the hill and shift cultivation of the Konds, Konda Dora and Jatapu Doras are remarkable. The jhola rice cultivation of the Bondos is most attractive. The Koyas rice fields, with provisions for fish management, are eye catching.

The poor agriculture by Sano Poraja, Ollar Gadaba, Bodo Poraja, Didayi and Kotia, Jatapu Dora, Parengas, and Kondhs in some regions is of concern. Gutob Gadaba and Sano Poraja have lost their rice lands to the Kolab Dam in recent years, leaving unyielding and uplands for agriculture. These areas are also facing shortage of fuel.

Bhattra

Plate 53: *Bhattra women selling pumpkin in market*

The Bhattras are also known as Bhottada, Bottada, Dholado. Stuart (1891) wrote, 'Bhattadas are a class of Oriya cultivators and labourers speaking Muria or Lucia, otherwise known as Basturia, a dialect of Oriya.' The language is classified as Indo-Aryan. Bell (1945) said, 'Bhottodas are the typical agricultural caste of Nowrangpur and north Jeypore' where they are found in greatest numbers. The tribe is also found in Kalahandi and Bastar districts of MP. As per census report of 1981, 117, 297 Bhattras can be found in M.P. and 247,710 in Orissa. Thurston and Rangachari (1909) cited Hayavardana Rao who said, 'the caste is divided into three endogamous divisions, viz.,

- Bodo or genuine Bhottadas
- Madhya, descendents of Bhottada men and non-Bhottada women
- Sano, descendents of Madhya men and non-Madhya women. Sano are also called Pitta Bhattra, nowadays.

The Bodos has various exogamous septs or *bamsa* or clans, which are as follows:

Kochimo	-	Tortoise
Bhag	-	Tiger
Goyi	-	Lizard (Varanus)
Nag	-	Cobra
Kukuro	-	Dog
Makado	-	Monkey
Cheli	-	Goat
Onka	-	Gourd

Bhattras are good agriculturists. They have settled cultivation and are living on plains of the districts. They cultivate various vegetables like gourds, pumpkins, beans, and yams, etc., at home. The villages are very neatly maintained, as are the houses. The houses generally have a large area around and a mud wall boundary. They are reputed for their cultivation skills under low moisture conditions during dry periods of the year. Rice and ragi is their staple food.

According to the recent census, 68.13 per cent are cultivators, 27.71 per cent are agricultural labourers and while 4.16 per cent are engaged in other occupations.

The Bhattras are of medium height and very well built. Their complexion varies from light skin to dark. Their heads are round, their noses are slightly brood and, on the whole, they are very attractive people. The women are attractive, equally well built and strong. Both men and women share agriculture work equally. They are a pleasant and friendly people.

Marriages are generally settled through negotiations. A price has to be paid for brides. Either spouse has the right to divorce and remarriage is allowed for both, widows and divorces. Marriages among cousins are common. In addition to their regular domestic activities, Bhattra women play an important role in agriculture, religious rituals and socio-activities.

A traditional community council headed by a Bhat Naik settles the intra-community disputes. Adultery, rape, divorce under traditional norms and theft are punishable offences and punishment is imposed in the form of a cash fine, social boycott and excommunication.

They have family and village deities and a pujari performs all the rituals. The rate of literacy is very low, with the female literacy rate being even lower. They claim to be superior in rank to Omnatya and Bhumia, with whom the men can dine, except in social functions. Bhattra women are forbidden from even drinking water from any other tribe, including the higher caste Brahmin and Karan houses.

The Bhattra tribe in Bastar are most progressive, and more and more children are being enrolled in schools. They are now in professions like medicine and engineering, etc. They accept modern technologies suitable to their life and living quickly. Similar acceptance of newer methods in agriculture and life is found in Koraput-Nabarangpur belt also.

They keep cattle, buffaloes and sheep. The ducks have a very special place in Bhattra life. The poultry is loved by all. Although they maintain goats when forest land is available, they somehow seem to prefer sheep. They have discontinued the practice of keeping pigs in their front yards. They believed that keeping pigs kept Mohammedan rulers away.

The Bhattras are highly skilled agriculturists, and very hard working people. Both men and women work in the fields. They have a large list of rice germplasm to their credit. They transplant rice. Rice cultivation is the main agriculture. They are also well known for being *attal* land cultivators. The cultivable land, in the beds of Indravati river and to some extent the Kolab river, is known as *attal* locally. They have developed least moisture cultivable crops. The *attal* brinjal, the *attal* tomato, which is slightly oval in shape, is specific to the place and tribe. They are skilled in the cultivation of sugar cane in *attals*. The castor, safflower, *masur* and channa, and many other legumes are cultivated in *attal*.

The backyard farming in the tribe is of great significance. In Bastar, Bhattras have 3-5 acre backyard farms for each household, but in Nabarangpur it is a small land patch. This cultivable land in the backyard is well protected by large timber pieces or mud walls. They grow creeper beans, pumpkin, gourds, bananas, several tuber crops, tomato, brinjal and other vegetable crops here. The periphery of the backyard has tamarind, mango, jackfruit and bamboo plants in most cases. As described earlier, the little gourd or tundri is very common in households, along with the Moringa tree.

The Bhattras are careful, hard working and progressive farmers. Their adoption of new technologies in agriculture is remarkable.

Kondh

The Kondhs are one of the largest tribal groups in Orissa. There are about 400,000 Kuvi speaking Kondhs distributed on the hills of northern undivided Koraput district. They have been variously spelled as Kond, Khond, Khand or Kondh, and Kandha. The word 'kondh' is derived from the word 'konda', which means hills in Telugu. MacPherson (1863), Campbell (1864), Bell (1945), Niggemeyer (1964), Benerjee (1969), Boal (1982) and Pattanaik (1982) have reported most aspects of the Kondh life and the way they live. The tribe can be classified under the following headings:

- Maliah Kondhs live in Phulbani and Ganjam districts. They speak Kuvi, a Dravidian language.
- Kuttia Kondhs dwell in remote hills in the extreme south-west of Phulbani district and continue westward up to Ambadola and Bissamcuttack of Rayagada.

Plate 54: Kondh family collecting wild mango

- Dongria Kondhs inhabit the steep hills of the Niyamgiri range of north-west Koraput district bordering Kalahandi.
- Kuvi Kondhs, whose villages spread over the hills and valleys of north Koraput district.
- Desia Kondhs are said to be migrant groups of Maliah Kondhs. They are influenced by the local population of non-tribes. They are found mainly in hills of Koraput, below the Eastern Ghats and the costal plains, or in costal Kalahandi, Bolangir and Dhenkanal districts.

Each group is further divided into a number of exogamous lineages like Jakasika, Hauka, Prska and Kadraka. Lineage names are prefixed to the names of individuals. Jani, Bishmagri, Pujari and Barika are surnames that Kondhs use.

In general it can be said that Kondhs are simple, straightforward people, who believe in community sharing traditions. They are a courageous and hard working people.

The Kondhs believe their spiritual traditions are the foundation of their survival and development. In the eighteenth century, McPherson gave a detailed account of the Khonds' religious activities, with

Plate 55: Kondh village front view in August

reference to *Meriah* (human sacrifice) investigation and prevention. The British made the first contact with Kondhs in 1835, whilst they were engaged in military operations for the reduction of the zamindari of Goomsur, in Ganjam district.

Elwin (1954) remarked that

> Deep in every Kond heart is the belief in the sacred and fertilizing power of human blood, and it is of course, for their practice of human sacrifices that they have become known throughout the anthropological world. There is extensive literature, ranging over a century, on the subject. Although the practice has long since been stopped officially, there can be little doubt that such sacrifices do sometimes occur, and a perusal of the stories collected will show how strongly Konds remain attached to them. Today, the usual substitute for a human being is a buffalo, which is called by the old name Meriah and in some of the remote villages; old human skulls are still used when the animal is offered to the Earth Goddess.

Boal (1982) wrote,

> Kondh relationship to their land is far more than economic. This will be seen more clearly regarding the central rite of the Meriah/Kedu (human/buffalo) sacrifice offered to the Earth Goddess. This and the great number of their agricultural rites concerned with fertility indicate both their outlook and their involvement in the land; for these rites not only incorporate petition to the deities but also express a sense of dependence on the land itself, linking each family almost ceremonially with its own fields and hill-plots.
>
> This attitude has a direct result upon land tenure procedures. For instance, when new ground is broken, even a small hill-plot, eggs as a fertility symbol are offered to the hill Gods and as year succeeds years, an ear of grain set aside from the previous crop is offered. This is distinct from offerings made to ancestor spirits who constantly accompany one on such work-occasions.

The Kondhs have been living in most unyielding hilly terrains for centuries. After the most arduous clearing of forest during 'podu' by all family members, away from the shadowing trees and exposed to sun, rain, light and air, these slope lands yield food through crops.

Very well built, the Kondh people take time to be friendly with outsiders. The women work as hard as men but do not take active part in group discussions. Although they stay as listeners, they keep away from the discussions.

The following are the deities that the Kondhs dwelling around Laxmipur block of Koraput district worship in different ceremonies of life:

Bhima Penu	-	Rain God
Idur Penu	-	New (seasonal) Fruit God
Jakri Penu	-	Village God
Dharti Mahaprabhu	-	Earth God
Darni Penu or Bizzi Penu	-	Lightning God
Saggu Penu	-	Thunder God
Surya Mahaprabhu	-	Sun God
Bondnui Penu	-	Forest God
Sooga Penu	-	Fountain God
Munda Penu	-	Tank God
Munda Penu	-	Stream God
Jama Penu	-	God of death
Idu Penu	-	House God
Sunali Penu	-	God of village boundaries
Lohra Penu	-	God of wars
Loda Penu	-	God of forest/hunting

The festivals and ceremonies of the Kondh life, and their relationship with livestock is detailed by Kornel (1999a).

Gutob Gadaba

Gadaba or Gadba is a primitive tribe. They live in Koraput and Kalahandi districts of Orissa, and in the Srikakulam and Visakhapatnam districts of Andhra Pradesh. They have been classified under the Austro-Asiatic group. Thurston and Rangachari (1909) first classified the Gadaba tribe under the following sections: Bodo or Gutob, Parenga, Otaro, Kaththiri or Kaththara, Kapu.

Thereafter Ramadas (1931), Furer-Haimendrof (1943), Bell (1945), Bhattacharya (1957), Thusu and Jha (1972) and Roychoudhury (1989)

Plate 56: Gutab Gadaba elderly women

classified the tribe in various ways. However, the Jeypore, Kundra and Koraput blocks mostly have Bodo or Gutob or Bhoie Gadaba group. Kornel (1999c) had conducted detailed studies on the Gutob Gadaba of Koraput.

Ramadas (1931) opined, 'In modern times, Gadabas take to all kinds of occupations. Most of them took agriculture, some lived by hunting and the others are traders.' At present, Gadabas are settled agriculturists in both the hill and plains of Koraput district. The success of agricultural operations depends on type of land, soil quality and available climate. On the whole Gadabas are among the best rice cultivators for ages, and have good water management skills in paddy fields. The Gadabas have a special ritual to collect and preserve the first harvested paddy as seeds. Both, Sabara tribes of Gunupur and Gadabas are ancient rice cultivators. With the origin of 1,800 varieties of paddy genetic resources, Koraput district is identified as one of the places of origin of rice in the world. The Gadabas have very rich ragi genotypes. It can be concluded that the Gutobs tribe are the foremost custodians of rice, ragi and other small millet germplasm.

The Bodo Gadaba are known as Gutob Gadaba in the hills, and are called 'Bohie Gadaba' in the plains of Jeypore because of their former position as planquin bearers to Jeypore zamindars.

The Gadabas are a cheerful and friendly people. Fair to dark in complexion, and of medium height, these well built, muscular men are fairly attractive. The Gadaba women are equally well built and strong. In old age, their faces are lined with numerous wrinkles.

With modern influences, the younger generation has lost the Gadaba dialects and has adopted the local Poraja or Desia dialects in Semiliguda and Jeypore. The Lamtaput group, however, is an exception.

The Gadaba women no longer wear large earrings or the Kerang around their heads. They have given up their homemade Kerang clothes and now dress like the others around them.

The Gadabas living on the hills still keep a few pigs, cattle, buffaloes, sheep, goat and poultry at home. The buffaloes are being used as beasts of burden in religious ceremonies, like Jankar and Gotar. Gadabas are well known in the world of anthropology for the Gotar ceremony. They eat beef and flesh of other livestock and poultry. The life and culture of the Bohie Gadabas, on the other hand, has been influenced by officials and various other people serving the zamindars of Jeypore, with whom they have been in contact for over a century. They do not eat beef any more, and feel offended if asked. They also do not keep pigs at home, and many have even stopped eating pig products. As tradition, however, some still continue to sacrifice purchased pigs in different ceremonies and festivals. They have even stopped the 'Gotar Ceremony', which was the centre of Gadaba culture and society. The younger generation of the Bohie Gadaba is more interested in the modern facilities, like radio, television, cycle, and motorcycle, etc.

According to the 1991 census, 42.69 per cent are returned as workers (65.78 per cent male and 19.82 per cent are females). A majority of these, 64.61 per cent, are cultivators and 29.06 per cent are agriculture labourers, 1.19 per cent live on forest produce collection and the remaining 5.14 per cent are engaged in other services.

Gadaba prefer adult marriage. Spouses are usually acquired by negotiation, elopement, intrusion and service. Monogamy is the usual form of marriage. The practice of paying for brides is common. Re-marriage and divorce are socially approved. The families are mostly nuclear in nature. The rule of inheritance is male succession and is through the eldest son.

Apart from carrying out domestic chores, the women participate in economic activities, particularly in agriculture. They are involved in the decision-making process in all spheres of family life.

The literacy rate is only 6.53 per cent which is very low compared to the average rate of literacy among Schedule Tribes, which is 13.96 per cent for the state.

The Gadabas are good rice cultivators and adopt new technology early. They are very competitive.

Bondo

The Bondos, who refer to themselves as 'Remo', are a small tribe of about 5,000 persons and are classified as Austro-Asiatic. May (1873) Henderson (1911), Thurston and Rangachari (1909), Furer-Haimendorf (1945), Elwin (1950), Das (1956) and Patnaik (1989) gave a good account of the Bondos. It took considerable effort and dedication from Elwin to prepare a detailed study of the tribe, which is known for its hostile behaviour.

The Bondos dwell around the Machkund river, in the Khoirput block of Malkangiri sub division of undivided Koraput district. They are distributed all over the hills.

Elwin (1950) wrote, 'The Bondo is a difficult person and his is a hard unyielding country. He has brought a higher degree of efficiency to the technique of keeping himself to himself.' They have, thus, retained their distinct socio-culture features.

The Bondos are distinguished by their distinct dress, particularly

Plate 57: Bondo women

the females, all of whom shave their heads. The women wear small, hand-woven, coloured skirts.

The Bondos are an endogamous community, divided into two totemic clans (bonso), namely Ontal (cobra snake) and Killo (tiger). These are further divided into a number of exogamous lineages (Kuda). These are Bodonaik, Challan, Dangra-Majhi, Dora, Jigri, Kirsani, Mandhara, Mundli and Sisa. The members of the Kuda are related through a common mythical descent and remain together under a common leader and magico-religious head.

The youth dormitories play an important role in the selection of wives. When a boy puts a brass ring on the forefinger of the girl of his choice and the girl accepts it, he gives his final consent, after which marriage negotiations get underway.

The wife is always much older than the husband. The Bondo are generally monogamous. Divorce is allowed from both sides, widow re-marriage is permissible. The nuclear type of family is the rule. The women play various roles in social, economic and religious affairs.

According to the 1981 census, 40.51 per cent of them are returned as workers (58.75 per cent males and 24.12 per cent females). Of them, 64.28 per cent were cultivators, followed by 33.17 per cent as agriculture labourers. Only 0.34 per cent were engaged in livestock, forestry, and hunting, etc., while the remaining 2.21 per cent were engaged in other services. The literacy rate is only 3.6 per cent.

The Bondos depend mainly on agriculture for food. They follow the traditional slash and burn cultivation on hills also. The change in the physical character of their environment has forced them from being food gatherers to cultivators. The hunting is limited to annual ceremonies. Roots and tubers are useful additions to their diet. Mango, tamarind and jackfruit trees are common in most villages and their fruit is eagerly collected, particularly during hot weather. Wild vegetables, bamboo-shoots and mushrooms are collected and eaten. They are very fond of red ants and date palm grub, which is roasted in fire before being eaten. They even fatten rats to eat and consider silkworms good for any swelling of the stomach.

They are non-vegetarians, and eat beef; their staple food is ragi and rice. They are very fond of fish, which they catch from rivers and streams. They eat carcasses of dead animals also.

Bondos cultivate rice on irrigated and terraced fields, and other crops on uplands. They also follow axe cultivation on the steep hill slopes.

The Bondos prepare seed beds in irrigated fields and rice is sown by February. Transplanting of rice is taken up in March, and by April the work in the fields is complete. This is a special feature of cultivation on hills because by June/July, when the monsoon rains arrive, the rice plant has taken root and holds on firmly to the ground and the soil. This prevents soil erosion. They reap the rice during December. From the time it is a seedling, till it is harvested, the rice plant takes between nine-eleven months. The Bondo show better skill in jhola rice cultivation. They do not broadcast seeds and do not grow upland rice. Transplanting rice is the rule.

The Bondos, like Kondhs, prepare bread out of mango-kernels in July and eat.

The Bondos keep cattle, goats, pigs and poultry. They do not milch cows but use goat milk in special ceremonies.

Elwin wrote, 'Work is shared by men and women, but the women labour more, and for longer hours in addition to work in the field or cleaning, they have to fetch water and cook and care for the children and whenever they are free they must be at the loom'.

Bondo Tribe Agriculture Calendar

The Bondo life is free, arduous but simple and, on the whole, happy. As with other hill dwellers, their lives are governed by the routine of the seasons, and the cycle of festivals and ceremonies. Their economic activities, social obligations and the religious activities combine in a coherent way. It has suited well to these isolated living people on hills and at the stage of development they are:

Month	Food production activity	Religious ceremonies of agriculture
January	Reaping and threshing of rice Threshing of ragi and other	*Sume-Gelirak* festival, the village priest ceremonially initiates the task small millets of cutting and cleaning podu land, in which the *Merer-Susum*, the first bean eating ceremony of the year takes place.

(Contd.)

February	Seed beds prepared in jhola lands and rice is sown. Ploughing begins for new clearings started on hills.	Ceremonial beddings. *Gunom* ceremonies.
March	Ploughing. Transplanting of rice-seedlings. Heavy field work.	
April	Work in open and irrigated fields over. - Forest clearing torched; fire set in Podu land.	*Giag-gige*, great festival begins from full moon of chait and lasts for one week. Mangoe, first fruits of the year eating It is now permissible to sow ragi.
May	Cleaning of Podu hills continue; unburnt wood collected and burnt Millets are sown.	
June	Rain begins - ploughing of land	Before the sowing,to promote fertility, fish and rice are offered in the fields. They broadcast mixture of several different kinds of *Eleusine corocana, Panicum milliare* and *Panicum italicum*, with *Sorghum vulgare* and *Penisetum typhoideum*. After the broadcasting they plant pulses of different kinds, including *Cajanus indicus* and *Dolichos biflorus*, cucumbers, gourds and castor plants.
July	Women weed the podu land	*Gersum-gige*, the feast of the new and remove sprouting shrubs, mango kernels and gourds, after which bread and gruel can be madefrom kernel-flour. *Gadar-gige* celebrated to protect cattle from diseases
August	Weeding in fields continues. The crops are regularly guarded, women collect bamboo shoots The early suana harvesting for food on a daily basis, starts from third week.	

(Contd.)

September	As above. *Alsi* (*Guizotia abyssinica*) sown.	First fruit feast of the maize crop.
October	Me dium duration millets are reaped, eaten at a ceremonial feast.	
November	Crops are guarded day and night, villages mostly deserted.	*Gewur sung* festival.
December	Reaping of rice, millets and crops in the hill cleanings.	New eating of siadi fruits and beans.

Plate 58: Bondo women transplanting paddy

Plate 59: Bondo women bartering agricultural produce with Dom women

Dongria Kondh

The Dongria Kondhs, are a group of Konds found in Bisamcuttack, Muniguda, Kalyan Singhpur of undivided Koraput district as well as in Bishwanathpur in Kalahandi district. They are hill dwellers and the area of their dwelling is called Niyamgiri hill range. It is between 1,200 to 4,900 ft above sea level. They speak Kuvi. The word 'dongar' means hills. They are basically shift cultivators and skilled fruit tree raisers also. They are altogether about 10,000 in number, of which about 6,493 are found in Rayagada area.

Plate 60: Dongria Kondh brother and sister

A good amount of work on Dongria Kondhs was taken up by Watts (1970), Nayak (1989), Das Patnaik et al. (1990), and Sahu (1992). Nayak (loc. cit.) has given a detailed account of Dongria Kondha, specially their feuds. In the early studies, there are several works on Kondhs as usual, but nothing specific on Dongrias.

The smallest social unit is family, and their lineage, can be traced to either Bismaghi Boinso or Mandal Boinso. Besides, there are clans, such as Wadaka, Kadraka, Pushika, Huika, Toika, Wangesika, Sikoka, Jakeshika, Nundruka, Nisika, Kutruka, Prashka, Meloka, Kandika, Tarika, Merika, Naraka, Palaka, Otaka, Prishka, Ongelika, Wanjelika and Piska. A particular clan from all the villages in a region forms a *mutha*. The bondage of lineage, *gotra* and *mutha* is strong, and plays an important role in all matters of their socio-cultural, economic and religious activities. In other words, the whole Dongria country, which is made up of hills, is divided into various *muthas* (*faṭri*), as per the social order. Thus, the *mutha* was the fundamental feature of social organization of the past and continues to be so even today.

The distinct feature among Dongrias is that both men and women keep long hairs, tied in a bun at the back. The men wear a loin cloth and a special scarf, whereas women wear a sari of three pieces. The men and women wear nose rings. That they earn their livelihood through horticulture is something specific to the Dongrias, as compared to other tribes of Koraput.

Women constitute the main workforce and, if he can afford it, a man is allowed to have two or three wives. The right to divorce is bestowed both the spouses. Widow re-marriage is allowed. The traditional council for social justice in Dongria villages is called *runda*. It is headed by *Jani*. The age-old Dongria practice of *Meriah*, where a *mutha* used to organize a human sacrifice has now been replaced with buffalo sacrifice.

The Jani worship the Earth goddess, while the pujari worship others. There's a peculiar relationship between Dongrias and Doms, a Scheduled Caste. The later is known to exploit the Dongrias in marketing and as moneylender. The Doms act as the messengers of the community in the outside world.

The hills are divided and demarcated for each *mutha*. The hills, are, therefore, treated as agricultural plots by individual families. These are hereditary and, if and when families break-up, these plots are also divided further.

The land in hill slopes is called *dongar* land. The land in the plains is called *gudia*. The kitchen garden behind, or on the side of the houses is called *badi.*

Their staple food, gruel is made from ragi and suana. It is called pej. They love rice, which is considered to be ceremonial food. Several varieties of small millets, pulses, and tubers, etc., are consumed. The *kandula* and *katting* (*Vigna umbellata*), along with Nagala roots, form bulk of the food. The fruits form a strong chain in food system. They smoke an indigenous tobacco *cheroot.* They eat the flesh of bovines, pigs and poultry, and have a particular liking for buffaloes. The pigeon is a delicacy. They even eat dead animals. They are very fond of dry, salted fish, which they get from the weekly markets. The men are fond of *mahua* liquor (*irpi kalu*) and Sago palm juice (*mada kalu*). They prepare one mild intoxicating beverage out of jackfruit. It is known as *panasamada*. The liquor has its ritual and social significance.

Well known as horticulturalists, they practice shift cultivation and are primarily hill cultivators. They grow mango, banana, jackfruit, pineapple, orange and lemon, etc. Bananas, oranges, pineapples and jackfruit grow luxuriantly in the hills. The off-season fruits varieties have gained much export value both in neighboring states and even to Delhi. A good Dongria farmer earns between 70 to 80 per cent of his total income from horticulture.

Their major income is from turmeric cultivation, followed by sale of jackfruit and banana. The pineapple is also a good source of income for them. These are their traditional crops.

Although recorded as government property, the forest land selected for cultivation is under the control of a particular clan in a locality. Forest produce supplements their income.

They grow cereal and legume crops for food. The mandia or ragi, kosla or suana, arka, ganthia, *katting* or rice bean, *kultha* or horse gram, *kandula* or *arhar*, kudijana or jowar, biri or black gram, jhata and turmeric are produced in the Dongar land.

The other crops are jada or castor, *makka* or maize and bailo, a type of bushy beans. The vegetable group is pumpkin, cucumber and *dhan maricha* or rice sized chillies. These are the traditional ones.

They do not grow brinjal, tomato, or okra. The *alsi* crop does not grow in their lands, is sadly expressed. The recently introduced crops and fruits are rice, wheat, onion, guava, *lichu*, papaya, sapota and illachi. Some improved varieties of the traditional crops have also been taken up.

Nowadays, there is hardly any game left in the region.

As per the tradition, the Dongria Kondhs do not go for wage earning to other villages. They have cooperative labour system (*punda-buti*). At the request of individual families the village people offer short period labour in sweden cultivation where they are provided with lunch and are paid few rupees at the end of the day. The employer family in turn attends to such request when called for.

The Kondhs in general are very skilled people. They are able to divert water springs found on the hills, through difficult terrain, for irrigation as well as for drinking water purpose.

The Dongrias are knowledgeable about agriculture and plant biodiversity.

The magico-religious practices and beliefs are associated with plantation in Dongria society.

The Dongria are a straight forward, but egoistic people. They usually do not pluck fruits for sale, which is considered to be a lowly person's job. The Doms take advantage of this attitude and on several occasions they mislead them at price level and hide the actual harvest.

Mango trees are commonly owned by the village and the funds collected by sale from such property go towards the village festival.

The government programmes for improving life, agriculture and horticulture, along with financial assistance and marketing mechanisms, have helped the Dongria tribe in recent years. The Dongria Kondhs are the most knowledgeable horticulturists on the hills. The religious rites expenditure falls heavy on families and they get indebt to moneylenders. On the other hand, a large non-agriculturist Dom population derive their earnings and living from them, is very sad.

THE CALENDAR OF AGRICULTURE PRODUCE, FOOD FROM FORESTS, AGRICULTURAL FESTIVAL AMONG DONGRIA KONDHS IN KORAPUT DISTRICT

Dongria month	English month	Fruits & Agr. crop harvest	Food from forest	Agricultural festivals
Pusha lengu	Dec.-Jan.	*Arhar* (*Kandula*)	Bhata kanda and Nangal kanda	Dongar puja, Ragi & other millet (suana) is worshipped
Magha lengu	Jan.-Feb.	Jackfruit	Siadi seeds green	Konga Parabu. it is the festival, where new Turmeric season arhar pods are roasted and all families consume *arhar* harvested. Meriah or Kondru Parabu: Meriah buffalo offer by mutha (once in 5 years)
Phagun lengu	Feb.-Mar.	Tamarind jackfruit sale continue	Salaph juice (intoxicant)	
Sitari lengu	Mar.-Apr.	Mango, jackfruit ripe nangla kanda	Jhadakuna kanda. Kundurukuna kanda	Bijchapa puja: Kosala, ragi, kangu, jhudunga, cucumber, *arhar*, kutting, pump kin seeds are worshipped for the next sowing season.
Basik lengu	Apr.-May	Ripe jackfruit and mango	–	The seed sowing on hills start with small offers by families
Lalindi lengu	May-June	Pineapple mango, jackfruit	–	Rakash puja: The devils worshipped on the road with chicken sale/buffalo/pigeon to ward off sad effect
Burel lengu	June-July	Mango, ripe jackfruit sale continues	–	–

(Contd.)

Srabana lengu	July-Aug.	Mango kernel	Bamboo shoots (one meal only), Banji mushroom from bamboo	
Bhadraba lengu	Aug.-Sep.	Mango kernel	Bamboo shoots, banji and basa mushrooms (khuda chatu)	Mandia rani puja: Chicken/Pigeon is offered to ragi
Dashara lengu	Sept.-Oct.	Mango kernel, early ragi, maize, orange, lemon,other small millets		–
Diali lengu	Oct.-Nov.	Orange, ragi pumpkin, jhudungo, katting, dakijona (sorghum) and tamarind vegetables available	–	Punapadi puja: Cattle, chicken are offered for good health and prosperity, including that of the livestock.
Pa lengu	Nov.-Dec.	Turmeric mature, ragi (rice), *arhar* and most vegetables available	–	Dongar puja and offer to harvest at the field site. Thanks giving ceremony

Plate 61: Dongria Kondh home garden with many economic plants

Plate 62: Hill cultivation with banana major crop

Plate 63: Hills slope turmeric cultivation

Koya

The Koya tribe, call themselves 'Koyiee'. Taylor (1862), Cain (1876), Thurston and Rangachari (1909), Grigson (1938), Bell (1945) have dealt with the Koya society in one another context. Kornel (2002) has studied the tribe extensively from 1970 and noted the socio-cultural changes since 2001. The Koyas live in the Malkangiri area of undivided Koraput district, as also in the neighbouring Andhra Pradesh and Madhya Pradesh, sometimes under different names. They speak a distinct Dravidian dialect.

Plate 64: Koya women in the village

The phratry recorded are Madkami, Madhi, Sodi, Padiami and Kawasi.

The Koyas are a good looking people. They are tall and have perfect physique. They have sharp noses and thin lips, shoulders are broad, and they have muscular bodies. The women are sleek and smooth, with well built leg muscles.

Plate 65: Koya rice cultivation and pisiciculture

The rice gruel and maize or ragi gruel forms the bulk of the food, along with available vegetables or animal products.

The agricultural festivals are *Mahua Pendule* or Mahua festival, the *Bija Pendum* or Chait Parab, *Chiquḍ Pendum* or *Chimi* or Bean festival and *Maska Pendum* or Mango puja. Kornel (2002) has described in detail the agricultural festivals of the Koyas.

They are mostly plain land slash and burn cultivators, and also rice farmers. Since 1960, there have been extensive changes in the locality where Koyas live. There was major influx of Bangladesh refugees under the Dandakaranya Project, of the government of India. The project had rehabilitated more than 12,000 refugee families in Malkangiri area. This lead to a large scale deforestation for agricultural land, to meet the requirement of refugee families. The land allocated per family was between 3 to 7 acres, besides homestead and village roads, etc. The project had also cleared forest land and offered it to several Koya families.

The following are the various types of agricultural lands:

Bariha Kuta is the low land where water can stand for eight months of the year; such plots are excellent for *Bariha wangi*, the long duration paddy.

The Koya farmer makes high standing bunds up to 4 ft., even for such purposes. The paddy varieties grow faster and keep above the standing water. Such land is prized because fish can also be harvested here. This is very special to Koya farmers of Malkangiri.

Uska beda lands are comparatively more upland than *Bariha kuta*. This type of land requires small earthen dams to retain water and serve the same purpose as *Bariha kuta*. *Isk beda* are those land where Sudul wangi, the early variety paddy grow. *Elka* are the uplands, where water does not stand.

Podu is otherwise known as the slash and burn cultivation. The first year of slash and burn land is called *Puna elka*. After first harvest it is called *Penta elka*.

The Koyas followed a primitive method of cultivating rice up to the 1970s. They plough the land once and sow the seeds on the same day. The paddy farms are hardly attended to and even weeds are not cleared. The cowdung manure was used at the homestead farm and not for paddy. They also continued with harvesting fish in the paddy fields. At the time of harvest, they would cut only the top of the paddy crop and burn the rest of the straw. This practice still continues.

Koya farmers believe that ash is a better fertilizer than manure. This may be due to their experiences in podu land ash effects.

Families grow beans, brinjals, pumpkin, gourds and mostly tobacco in their backyard, which is heavily manured with the cow dung manure. Many families maintain a small vegetable garden near the village field, which also has small thatched farm house. The families generally rest here after a hard day's work and even cook here. They generally grow vegetables here.

They grow ragi, millet, beans, jhudungo, pesli, or pili pesra, *kulthi*, etc. In the early days, they used to barter these for salt, etc. Somehow this practice still continues to some extent.

The *mahua* tree is the very basis of the Koya economy. The flowers of *mahua* are carefully collected, and the fruit called 'tolo', is used for oil extraction by indigenous methods. The tolo oil is used for both cooking and hair dressing. The surplus oil is sold.

They prepare beverages of *mahua*, rice, mango, and tamarind, etc., and collect fermented juice from palm and sago palm trees.

The animal husbandry among Koyas is specialized. They rear cattle, pigs, goats and poultry.

The Koyas of eastern Malkangiri do not milk cows and do not drink. They have a large number of cattle at home and in the village. Cattle are considered a symbol of wealth. Mohapatra (1992) opined that Koya economy depends on their cattle. The animal husbandry of Koya tribe has been described in detail by Kornel loc. cit.

The cattle are kept out doors, in the open, all through the year. In recent times due to non-availability of grazing opportunities, the number of cattle has reduced.

The Koya tribe is reputed for breeding goats in large flocks (Kornel, loc. cit.). They are known to have a special preference for raw pork. The tribe keeps dried and smoked animal muscle pieces in bamboo tubes, to entertain guests at home.

The Koya tribe of Malkangir, are in close contact with Bangladeshi settlers, particularly, in the Potteru irrigation project area and are fast changing towards an improved agriculture system oriented towards market economy.

Sabara

Sabara is a known ancient tribe of India. It has been variously spelt as Saora, Soara, Sabar, Shabar, Sour, Sar, Sayar, Suir, Sahar, Soria, Swiri, Siwir, Surris and Saura, etc. They are Astro-Asiatic in origin.

They also inhabit in Koraput-Ganjam (Gunupur and Paralakhemundi region) and western Orissa. This work deals mainly with the Sabaras of Gunupur region. Thurston and Rangachari (1909), Roy (1927), Ramamurti (1931), Sitapati (1938), Dasgupta (1938), Elwin (1955), Patnaik (1963), Panda (1965), Mohapatro (1965), Watts (1970), and Bhupinder Singh (1984) have conducted detailed studies about the tribe and written on various aspects of their lives.

Plate 66: Sabars family in agriculture operation

Elwin (1955) gave a good account of the Sabara tribe of Koraput in his study. We have taken the following classification of Sabaras from his studies:

- Lombo Lanjhia are the Hill Sabaras.
- The Jati Sabaras are a branch of Hill Sabaras, who do not eat beef.
- Arsi Sabaras wear cloth and are Hill Sabaras.
- Jadu Sabaras live on the hills of Serango. They claim equality with the Jati Sabaras, but the latter disagree.
- Kindal or Takala Sabaras are basket makers.
- Kumbit Saoras are potters.

The Sabaras are ancient agriculturists. Elwin (loc. cit.). continued, 'To the Saora every part of the agricultural process is a religious act; his field is a temple, his hill cleansing a shrine. They cultivate homestead land, undulated rain fed lands at the base of the hills, terraced lands as well as swedens in the hills.'

Elwin further wrote, 'For the Saoras cultivation is something more than just a means of getting food to eat. It is a whole way of life. It is arduous, but it is happy. It winds an entire family in a single task.'

The Sabara women are given the right to plough their land by bullocks. Except the Muria women of Bastar, most tribes in the Koraput district do not allow such liberty to their women.

The homestead land is cultivated with maize, chilly, tobacco, pumpkin, gourd, bean, brinjal, papaya, banana, lemon and orange and small millets of various kinds.

The other type of Sabara cultivātion is the slash and burn or shift cultivation (*Bagada chasa*). It is also called axe cultivation, and has been described separately. According to Elwin, this is not as important to the Sabaras as it is to the Baiga and Maria tribes. It is important to the hill dwelling Kondhs of Koraput and Pengos of Kashipur.

The hills around the village were traditionally distributed on the basis of 'Birinda' or extended families. When they are cleaning the forest for shift cultivation, they do not cut *barada*, *arandah* and *karki,* which provide edible leaves. A creeper called *samatila* also has edible leaves (Patnaik 1989).

The cooperative labour that is followed by Bondo, Dongria Kondh and others is kown as *Ansir* to Sabaras and is utilized in slashes and burn cultivation.

They follow certain rules in cultivating cereal and legumes on hills. They prefer cereal or pseudocereal crops in the first year, like *jhudunga* or *burbudi*. The *jona* (sorghum) and *arhar* are not grown in equal proportion in the mixed cropping stands. This is to avoid competition of growth and nutrients. In the next year, a reverse trend of species is followed. In later years, however, when fertility is declining, they exclusively cultivate the legumes and leave it fallow.

The terraced field, where water flows round the year, is called *sarroba*. It is only meant for rice cultivation. Many Sabaras raise rice crop for profit, and grow two crops per year. The main rice crop is sown in June, while the second one is in January. The latter is called 'mango-rice'. Generally, rice is grown by transplanting, but sprouted seed technique is also employed.

Elwin wrote,

> These terraces are works of no small engineering skill. Built right up the beds of mountain streams, rising for many hundreds of feet from the depths of the valleys to the hill tops, they occupy every inch of ground available, sometimes only a few feet broad, they creep round the hill sides. They drop from terrace to terrace may be as deep as fifteen feet, yet so carefully are the stones facing built up and so flat are the platforms that no soil is carried over by the water that trickles from level to level down to the plains below. Ingenious methods of irrigation have been devised to control the flooding of the fields.

The Sabaras' name for months are not based on the Hindu lunar system but on their agricultural operations. This is a distinct and important deviations unique to the Sabaras. At the same time, this also suggests the strong agricultural influence embedded in the life and culture of the Sabaras. Patnaik (loc. cit.) has given the following in clear terms.

The religious rites on different occasions of harvest, as listed below, indicates the importance of the crop to Sabaras.

Sabara month	English equivalent	Explanation of terms	Agricultural operation, after which the month is denoted
Galanbur Gae	January	Galan (*arhar*) Bur (harvesting) Gae (month)	Harvesting of *arhar* & *jona*
Niva Gae	February	Niva (fire wood)	Carrying fire wood from swedens to home.
Ariamdur Gae	March	Arimadur	Forest cleaning and firing

		(firing)	the slashed trees, bushes in the hills.
Gusang Gae	April	Gusang (dibbling)	Dibbling of *arhar* seeds in the field.
Lala Gae	May	Lala (hoeing)	Hoeing the swedens.
Pujing Gae	June	Pujing (weeding)	Weeding in swedens.
Gusai Gae	July	Gusai (transplanting)	Transplanting of ragi.
Gusar Gae	August	Gusar (transplanting paddy)	Transplanting of paddy.
Tisar Gae	September	Tisar (weeding)	Weeding in terraced fields.
Gapal Gae	October	Gapal (weeding)	Weeding of fields.
Gajing Gae	November	Gajing (cutting grasses and forest cleaning)	Forest cleaning.
Timbir Gae	December	Timbir (forest cleaning, as above)	

It is strictly taboo for anyone to eat part of a crop until ceremonies have been performed by the elders on behalf of the whole village.

Abbanadur: In April or early May, the *mahua* trees are covered with fresh new leaves and flowers begin to open. Before the *mahua* flower is collected, eaten or distilled into liquor, a festival, Abbanadur, has to be observed by all.

Bunyanadur: It is held in August for *Panicum italicum* and other small millets.

Enlud anadur: This festival is held at the end of rains for the new cucumbers (*Cucumis sativus*). It is only followed by Ganjam Sabaras.

Gangugayanadur: This is for sweet potato, *Dioscorea alata*, but is extended to all kinds of wild roots. It is observed in July.

Kondemanadur: The festival is for coarse Kondem rice (*Echinochloa colona*), which is supposed to be preferred by the gods, for it is an older crop than the finer rice that is now cultivated (Elwin, 1955). It falls in the end of September.

Kurojandur: The festival is for the small bulrush millet, *Holcus spicatus/Pennisetum typhoides*, which occurs in October.

Osanadur: The festival is for an important millet, *Eleusine corosane*, which is held in October.

Pujjingadur: It is a small rite performed in July or August, at the time of heavy weeding of fields.

Rogonadur: It is a major festival for the harvesting of the red gram, *Cajanus indicus*, which is carried out in December or January, according to the yield of the harvest.

Tankunadar: This festival is for the stones of the mango. They dig stones of mangoes in sun by sprinkling little salt.

These festivals give the impression that *mahua* (*Bassia latifolia*), small millet (*P. italicum*), cucumbers (*Cucumis sativus*), sweet potato (*Dioscorea alata*), coarse kondem rice, bulrush millet (*Heleus spicatus)*, ragi (*Eleusine corocans*), redgram (*Cajanus indica*), and mango are the most favoured crops. This may be due to the fact that they are 'life saving crops'.

The weeding of field crops is another important occasion that can be seen from here.

The agriculture practised by Sabaras is complex, and quite advanced among the hill dwellers. They have life saviour crops, famine saviour fruits and crops and crops for food and sale. It is a mixed species of crops. There are more than 35 species of plants in the system. They have a broad spectrum of genetic resources in mixed spice plant agriculture. The land cultivated for food is sloppy, steep, plain and uneven. It is rain dependent, and only some parts have perennial flowing water for irrigation. It contains legumes, cereals, vegetable crops, fruits, oil crops and even medicinal plants. The seed rate of plant species, spacing between plants and species, sequence and time of sowing, and the plan of placing different species in different parts is amazing. The plan contains provisions to avoid plant competition for growth between plants. The biocontrol measures through various species as interculture is also within the system. This system has assured food supply to the family.

The plant protection: From the time the seed is taken to be sown in the fields, they watch over it with loving, anxious care. And when the plants start growing, their long struggle against unseen enemies begins. They perform minor rituals to protect the plants and stimulate their fertility. This period is filled with rituals and offerings.

Plate 67: After harvest, during summer, the fields with casava papays and banana crops

Plate 68: Banana standing crop in the shift cultivation after total harvest

The Harvesting festival: Elwin writes, 'A final discipline controls the farmer eager to enjoy the fruits of his soil. It is strictly taboo for any one to eat part of a crop until ceremonies have been performed by the elders on behalf of the whole village. The enjoyment of the harvest must never degenerate into a merely individual triumph over nature.'

They recognize the gods as the real landlords of the fields and clearings, to whom tribute must be paid.

Sabara Hill Cultivation

Hill Sabara families own about 1 to 2 ha of sloppy land on hills; of which 25 per cent remain under slash and burn cultivation and another 75 per cent are terraced land. Terraced land are for cultivation of cereal and legume mixed crop, whereas the borders are for growing fruit trees and tuber crops.

The described cultivation is the general rain-fed *kharif.* The cereals are kangu (*Setaria italica*), suana (*Echinochloa frumentacea*), *jona* (*Sorghum vulgare*) and ghantia. This includes the early and late varities of jhudunga. The early variety is called Chotodi while the late variety is called Holodi. Early maturing rice is cultivated in about 20 per cent of the lands. It has mixed ragi and jhudunga. Rice is transplanted from nurseries to Sarroba lands. A small narrow walking path leads to the hill, where a small hut, sufficient for the family, is constructed. This provided a good view of all the farmland. The natural forest on the top of the hill remain intact.

Families usually leave an area of about 6 ft. in front of the house, and 4 ft. around it, to plant papaya, banana, cassava, tundri, pumpkins, maize, jhudunga, sapuripanasa, kaunria sago, etc. All homes generally have two hens and a dog.

On the slope of the hill that leads to the house, again, kangu, jona, maize, ghantia, jhudunga and *kandula* are usually planted. On either side of the drains that are meant to carry excess rain water, both in homes and on the hills, they plant the kaunria bushes. Its tender leaves grow all round the year, and are cooked for food. These are slightly sour in taste. The jada or castor plant is planted at the periphery, and also between large stones.

A local variety of medium-sized bananas, called 'Deshi Kadali' are planted. Yellow when ripe, these bananas are sweet, tasty and fragrant.

They fruit all through the year. It is planted around the bunds, preferably close to the nallas.

The cassava is called 'Kathakonda', and also 'Semulikonda'. It can stay up to nine months before harvesting.

The following crops are harvested against the months denoted

- Kangu flowers in July and is harvested at the end of July-August for family food. It is harvested by removing the ears of the crop with a knife.
- Suana is also harvested by July.
- Jona or Sorghum is ready for harvest in October-November.
- Bodo Suana is harvested in October-November.
- Ghantia (*Pennisetum typhoideum*) is harvested in October-November.
- Dongarani is harvested in October-November.
- Chotodi Jhudunga is harvested in October-November.
- Mandia Bogodo is harvested in October-November.
- Kandula fruits by November-December. Its green pods are used for food.
- Jada (kala) is harvested in February.
- Early maturing paddy (Kunda Dhano) is harvested by September.
- Desi Khambalu is harvested in January and February. It once in two years.
- Kadali Desi fruits throughout the year.
- Pumpkin matures by July but is sown during April.
- Long duration Jhudunga is named Holdi variety and it yields beans for about 8 months.
- Jhudunga Jhota variety is the early flowering type.
- Bhendi (Lady's finger): The seven leaf variety fruits round the year.
- Ghantia : It fruits in July and lasts for 30 to 40 days. It is sown in April.

- Tundri is called Taruda here and fruits round the year. It is planted in April.
- Desi maize — its cubs are harvested by July end and continue for ten days.
- Papaya is called Desi variety and starts flowering after six months of plantation and continues fruiting for 4 to 5 years.
- Maricha (chilli), a local type, it bears fruit round the year.
- Kaunria saga, a leafy vegetable, yields leaves all the year round. It requires moisture.
- The sweet potato is harvested in October-November.
- Lemon fruits in summer. Oranges are common and Sapuripanasa or Pineapple is available by October.

Every two or three years the Sabaras sow turmeric and ginger in a part of the swedens. They also sow the oil crop *Guizotia abyssinica* on the hills.

A row of castor is planted along the path through the maize field, on which the livestock walk, to protect the crop.

This system has cereals, legumes, tubers, vegetables, fruits, fibre crops, oil seed crop and medicinal plants in it. The Sabaras cultivate more than 35 varieties of plants for food. There was more than 23 varieties of cultivated plants recordes in one of my studies in a Sabara family.

The crops range from the early maturing type to the late maturing type, along with perennial fruits and vegetables. The cereal, legume base is best utilised as a mixed crop. They do not use insecticides or weedicide, and even chemical fertilizer application. There is assured food supply to the family under any adverse circumstances. This farming system is time tested and has inbuilt assurance. The farmer is highly skilled in maintaining plants and, depending on their individual requirements, allocating appropriate space and location. They are knowledgeable about various aspects of farming, like maintenance of seeds, sowing different seeds in sequence and in appropriate climate and soil conditions, and in planting materials for the next season.

Poraja

They are variously known as Parja, Paroja, Porja, Poroja and Parji. The word 'praja' in Sanskrit means the subjects of rajas or kings. The tribal legends also suggest the same.

The Poraja are not a compact community, but a conglomeration of several endogamous divisions. Their language varies according to locality (Thurston and Rangachari, 1909).

The main concentration of Poraja is in the district of Koraput and Kalahandi, with the majority of them being found in the Koraput subdivision as well as Nabarangpur-Kashipur regions.

The Poraja community was classified by Thurston and Rangachari (loc. cit.), followed by Bell (1945). They have classified some endogamous sections in Poraja, called Barang Jhodia, Pengu, Khondi or Kondi Poraja, Parengi, Bonda or Nangla Poraja, Tagara and Dhur Poraja.

Here two branches of Poraja, namely Bodo Poraja and Sano Poraja have been described.

Bodo Poraja

The Bodo Poraja tribe is a small group living in villages around Jeypore and the Koraput subdivision of undivided Koraput district. The main concentration of the tribe is in Koraput subdivision, mainly in the hills around Kakrigumma and Dasmantapur. The synoname to the Bodo Poraja tribe are Bod Sodia, Sodia and Sodabisia. To the neighbouring tribes, they are also known as non-beef eating Poraja.

Thurston and Rangachari (loc. cit.) and Ramadas (1936) have described Bodo Poraja under the heading of Poraja of Jeypore. The Bodo Poraja have clans or Banso, which are:

Nag	Cobra snake
Khora	Sun
Bhalu	Bear
Hanu	Black faced langur monkey

Bagh	Tiger
Phulu	'Niali' flower
Jani	Priest (principal)
Disari	Priest

Some of the Bodo Poraja reject this clan classification and maintain that Nag, Hanu, Bagh and Phulu are only the clans. The rest, they say, are titles of hereditary family positions and places of origin.

Plate 69: Bodo Poraja

Marriage between cousins is the general rule and usually takes place when the two concerned people have reached adulthood. Widow marriage is allowed, so also divorce. They live in nuclear family structures.

The Bodo Porajas are having distinct Dravidian features. They have very dark complexions, sharp facial features, short structures, but well built. They are shy and timid by nature. They say, 'We Bodo Poraja run after tigers, but we are afraid of human beings'. They prefer to live secluded lives. They are a simple people singular in faith and belief and honest.

They speak 'desia', with an accent all their own. The women of the tribe can be identified by their distinct hair-do. They also wear small rings around the rims of their ears, which is peculiar to the tribe.

The participation of women in village discussions is limited. They stay out in groups at a distance, when men are involved in discussions.

The Bodo Poraja women cannot accept food and water from a person of any other caste, not even from the high caste Hindus. The men, however, can accept it from people belonging to the Rana, Gouda, Matia and Paika castes only but not from others.

Agriculture is the primary occupation of the Bodo Poraja. They practise shift cultivation and settled agriculture. Their agriculture production is facing degeneration. The fertility of the red, loomy soil in the region is poor and has high acidic pH. They grow ragi, suana, rice, *kandula* and *alsi* in the fields, and small quantity of semi beans, pumpkin, maize and vegetables in the family garden.

They rear cattle, buffalo, sheep, goat, pigs and poultry at home. They have numerous agricultural festivals every year. These have been described by Kornel (1999a). The main festivals and ceremonies are Mandia Rani puja (ragi-queen puja), Bondapan puja of rice fields, new semi fry in bhogi, new *kandula* fry in jankar, and the common khola or harvest ceremony.

In Chait Parab, they worship new seeds for the coming agricultural season. In the Bandapan puja, they erect twigs of kendu, Bonobhalia, Deobadni, Garadphula, Bonodhungia (forest tobacco) and Jalandrimal in the centre of rice fields to save crops from pests.

During Dussehara, they observe Johnijatra, after which the cattle can be used for paddy thrashing. The Diali Parab is celebrated by feeding the livestock at home. The new harvest rice is cooked with vegetables and a little wine and fish. The family livestock is offered the same food in new bamboo baskets. It is a thanks giving ceremony to livestock.

The families celebrate Nuakhia Parab (new year eating) in Bodh month. The new paddy and ragi panicles, pumpkin tendrils, aloes, tagrakanda and kudaie leaves from the forest and tamarind are worshipped during the puja. After this, the villagers eat newly harvested food crops.

Sano Poraja

The Sano Poraja tribe dwells in Koraput and Jeypore sub-divisions of undivided Koraput district. The concentration of the tribe is well marked in the Koraput subdivision. They speak 'Parji' the dialectic, which is also known as 'desia', a very close form of Oriya. Ramadas (1936). Mohanty (1997), Ulaka (1976) and Mohanty (1990) have reported on Porajas of Koraput.

Plate 70: Sano Poraja family preparing pendum drinks

The Sano Porajas are considered to be lower in social status by Bodo Porajas of Kakrigumma locality. They are also known as Barang and Jhodia in some localities.

The Sano Porajas are a peace loving people. Dance, music and songs are vital in their lives. On the whole, they are a very pleasant, docile and hospitable people. They prefer to live a secluded community life. This is evident from the fact that, although they have been able to sustain their lives and culture without being influenced much. What is most surprising is that, unlike other tribe of the area, they are not very competitive and have made no effort to claim the benefits of the development programmes around them. They do not blame anyone either. They are a simple, innocent people and very honest.

The Sano Poraja have various clans, or Kuda, and they are: Jani, Bhoie, Mudli, Kirsani and Narangia.

The Karangia, Bodo-Nayak, Nayak, Pujari claim to be Kudas or clans; but actually the titles have been derived fom their hereditary official positions in villages.

The ladies have some say in decision-making at home and also in matter relating to village welfare. As observed during village meetings, this is true of the Bhumias but not so much about the Gadabas. The older women enjoy far greater freedom to interact with visiting officials. The Sano Porajas are not good communicators. This has been observed during discussions.

It appears that Sano Porajas are consuming more non-vegetarian food than the Bhumias of the plains. They eat buffalo, as also old or dead cattle. They are fond of pigs, poultry and small ruminants.

Diseases like malaria are common among Porajas, as is sickle cell anemia. They have comparatively poor health and hygine.

The dead livestock is used as food, and the skins are sold to Harijans. They do to not sell cattle/buffalo hide to anyone except the Harijans.

The festivals and ceremonies of Sano Poraja tribe are detailed by Kornel (1999a).

The soil fertility around their villages is very poor. It is highly acidic pH. The water retention capacity of the red loamy soil is poor. The soil contains high concentration of aluminium and iron, but is deficient in boron, molybedinum and zinc. The phosphorous is also very low.

Agriculture depends on rain. The land and hillock farming give poor yield. Few families who have jhola land, cultivate rice and are better off as compared to the village members. Their staple food is ragi. They grow ragi, suana, jona and rice. The backyard has some semi, pumpkin creepers and other vegetable and if possible few plants of maize. As short-term oil crop, *alsi* is grown extensively on hillocks.

It can be concluded that Poraja agriculture is primitive and poor. The soil and moisture factors are responsible.

The family members, including women, work hard in the fields to earn food for the family. They are very traditional in their outlook, unable to accept new technology and ideas. Being poor, they lead a hard life on the hills. If the crops are not regularly rotated in the field, they are prone to attacks from while ants. Environmental degradation has had an effect on their lives. With the forest having already disappeared, they make their living as fuel collectors, selling charcoal in the town nearby.

Bhumia

The Bhumia tribe is mainly concentrated in the Jeypore subdivision, mostly in Jeypore, Kundra, Boipariguda, and Ramgiri blocks, and also in Nabarangpur sub division. Their population is over one lakh.

The Bhumia divide themselves into Bodo and Sano groups. The Sano Bhumia are also known as Buri Bhumia or Matya who mostly live in Ramgiri area, on the other side of the Kolab river. Nowadays, restricted numbers of marriage between both Sano and Bodo Bhumias are taking place, which was forbidden some years back.

The various clans or *bonso* of the Bodo Bhumias are:

Nag – Cobra snake
Bagh – Tiger and
Surya – Sun

Plate 71: Bhumia tribe young women celebrating chait parab

Ordinarily, in one village, the family of one clan resides. Inter-clan marriages are allowed but not intra-clan marriages.

The Bhumia are tall dark complexioned thinly built people, considered handsome men and beautiful women by neighbouring tribes. They are hard working and settled agriculturists. On the whole, they are an intelligent and wise people, possessing calm behaviour. They show a very highly collective and disciplined behaviour in their society and in life, generally.

They do well in dry land cultivation. Ragi and rice are their staple food. They are good in rice farming. They grow *kandula* and *khedjona* (sorghum) in their fields. The backyard farming is well maintained by women. Fruit trees form a large part of living, particularly jackfruit.

They prefer sheep, poultry and duck, and also rear cattle and buffaloes.

They quickly follow and adjust to new developments in agriculture. They are quite particular about personal hygiene, as well as that of their village.

Tribal Magico-Religious Rites in Agriculture

The pressure of a prospective production has been the worry, at both the family and agriculture level. It is not surprising to find villagers busy with one ritual or the other, everyday. The whole process of agriculture production and food collection events go through several magico-religious rites, which are sometimes celebrated as festivals. This starts from forest clearance and continues for land preparation, seed sowing, safeguard promises of crops until threshing floor puja after good harvest. A vital part of the tribal culture is, celebration of new year crops and fruits before the community starts eating. This is global in tribal world. A detailed festivals and ceremonies calendar related to livestock and poultry offers, has been reported (Kornel, 1999a) for ten tribes of Koraput district.

Plate 72: Bhumia tribe performing magico-religious agriculture rites

Elwin (1955) conducted a detailed and exhaustive study about the religious aspects of the Sabara tribe of Koraput, covering even their magico-religious beliefs about agriculture and food collection. Several other authors have also generally dealt on the subject in their works.

Elwin (1950) detailed the agriculture and general ceremonies of Bondo tribe. They have complex magico-religious rites in seed conservation, fertility and production. Almost all the animals they have at home have been promised as offering to one deity or another. This made it difficult for Elwin to purchase poultry from the Bondo village during his stay there. The works of Elwin (1947), on the Muria tribe, and Grigson (1938), on the Maria tribes of Bastar, give total account of agriculture related ceremonies and the inner feelings of the tribe.

The magico-religious effect on soil fertility, particularly the production and colour of turmeric among the Kondhs of Orissa has been exhaustively reported by former British officials. The prevalence of human sacrifice, every three years in a cluster of villages, was rampant up to 1863 in Koraput, Ganjam and Phulbani districts. This ceremony was called *Meriha*, which still exists as 'Buffalo sacrifice'.

Their magico-religious relationship with agriculture production is still a complex subject, particularly among the isolated, and primitive tribes of Koraput and Bastar. This needs to be considered seriously, especially when development plans are prepared in these spheres.

Threshing Floor

The threshing floor is believed to be the most important place, which determines the outcome of the harvest to the family. There are a lot of taboos attached to this site selection, cleaning, and painting it with cow-dung paste. After a brief ceremony including sacrifice of a small animal, even may be a rat, the farmers begin crop harvest. The Gadabas do not allow women into the threshing floor during rice threshing. Most tribes do not allow any person other than their family male members to enter the threshing floor. Some tribes allow women to transport the harvest material through headload. However, no tribe allows women to fill the basket or bags with grains, and carry home from the threshing floor.

Therefore, it is very important for outsiders, like scientists hoping to collect seed sample, to seek prior approval to enter the threshing floor. Serious feuds and displeasure can appear under such circumstances.

A great harvest is enjoyed with a ceremonial offering like pigs (if the tribe likes pigs), or a rat, which looks like a pig. The pig is the symbol of fertility and prolificacy to the tribe. Some sacrifice fowl.

Some tribes perform special ceremonies for storing seeds. This is a serious affair for the family. This has been described elsewhere (Kornel, 1999a).

While ragi and millets are thrashed with sticks, paddy threshing is done by making the cattle and buffaloes walk over it. The women are allowed in ragi threshing.

After a new crop has been harvested, and before its produce being enjoyed in a meal, each family of the village participates in a community ceremony.

Seed Germination Bali Puja

Seed germination in Bali puja is followed in tribals like Bhumia, Bhattra, Konda Poraja, Pengo and others. This is collective function in which villagers from many villages gather at one point. Each family, or clan, bring seeds and bamboo packets to the Bali puja. This lasts for nine days. After the germination and luxuriant growth of the said seeds, final celebration is performed.

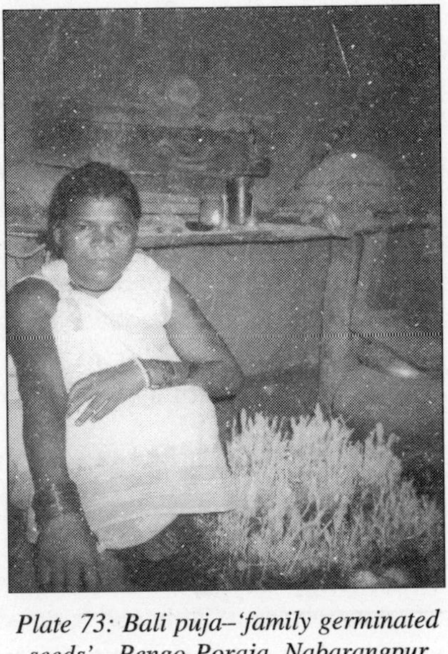

Plate 73: Bali puja–'family germinated seeds'—Pengo Poraja, Nabarangpur

Some tribes, like Bhattra, perform the Bali puja once in every twelve years. It is an elaborate function and lasts for four months. The process is followed as described earlier.

Before sowing the first seeds of the season, the family consults the Disari for an auspicious day and time. They even follow particular direction to face while sowing seeds. The men only sow seeds in the fields. They start with *dhupa* and offer flowers. In some tribes eggs, which are considered to be symbols of fertility, are also offered.

The Tribal Cooperative Agriculture Labour

The primacy of reproduction of the productive unit (the domestic household) over the actual production the unit engages in, remains a characteristic feature of the traditional economy (Nanda, 1994).

Nanda further wrote with reference to cooperative labour that the Bondo highlanders followed, with the patrilineal clan base expressed solidarity, which may be called a 'mechanical' solidarity reactivates at the subsistence production.

There is a Bondo local system called *Odja*, wherein an individual household that requires assistance takes the initiative to organize a labour assembly. This cooperative labour organization, offering traditional labour aid, has remained viable primarily because it helps accomplish jobs that require large number of persons, or which must be finished quickly. The Sabara call this *Ansir*.

The Dongria Kondhs have a similar system, with a strong clan as well as 'Mutha' base. The 'mutha' is a cluster of villages. Among the Koyas, the same institution is based mostly on the solidarity of village membership.

These include cutting down of forest cover before the monsoon sets in, to broadcast seeds, erecting fences before the monsoon arrives, harvesting in time so that grains do not scatter, uprooting and carrying a thatched watch house tower to a new location; whereas among Koyas, it is the erection of damaged bunds due to flooding in the paddy fields, or the construction of strong bunds in paddy fields before the arrival of monsoon rains.

This customary practice of labour assistance is based on mutual obligation and symmetrical reciprocity.

A senior male member of the household decides the day when assistance is needed. The selected person are orally informed a few days in advance and invited for the particular work. Those who work are paid both in cash or kind. Among Bondos, plenty of millet gruel, along with *mahua* liquor or palm juice, is served on the day of *Odja*. Among Koyas, the favourite is a pig with rice gruel and some wine or beverages. The owner also owes, in return, one day of *Odja* labour to each of the labour participants. Adults return the favour on appropriate occasions. The norms remain flexible to the compulsions of production.

In cases of very needy families, neighbours and kith and kin volunteer to help, and even bring their own food and drink. Similarly, among the Koyas, families without cattle are gifted few cows, which they return after few years, when they have enough to breed.

It is a customary practice in tribal villages to lend cattle for an agricultural task in turn for a similar favour. This is mostly seen at the threshing floor.

The Bondo children play *Odja* during childhood, and then assist the elders in the fields. The practice helps instill pride, pleasure, esteem and honour of cooperative labour among the Bondos. The *Odja* also strengthens social relationships of the community. A similar system of cooperative labour, prevalent and is called *Punda-bhuti* is common among Dongria Kondhs.

There is another type of labour distribution which is extended to the common land that belongs to the clan as a whole. In such *Odja* there are no labour debts incurred nor any credits earned. The clan allegiance provides collective strength, and is called upon under extraordinary situations. Similarly, in a modified line in the plains of Koraput, the village forest land is provided protection through paid watch, or by each household in the village, which keeps watch on a rotational basis. This practice has originated recently, due to acute shortage of commodities from the fast diminishing forests.

The contractual wage labour is widely known as *cooli* or *bhuti* or *cooli bhuti*. This has largely replaced the age-old *goti* system and other forms of labour. Such labour is employed in harvesting, transplanting, weeding, threshing and other agricultural work. In such cases, the tribal owner works beside the employed labourers, or *bhuti*. Here, work is done in return for cash. It is also a quantitative shift in production (Nanda, loc. cit.).

The *bhuti* is also paid in advance to the person concerned, when there is acute shortage of food or the family has to meet additional requirements.

Nayak (1989) has described similar labour cooperations (*punda*) among the Dongria Kondhs. There are mainly four types of *punda*, which are *Saha* or *Satra-punda*, *Dutaru-punda*, *Dhangadu-punda* and *Daasi-punda*.

The *Satra-punda* is a large group structure with families of the village, or satellite village. Here the Dombs too participate. The person who requests help of *Satra-punda* pays Rs.10 to the village fund, and places his request in presence of village council at *Kudi*, early in the morning. He has to offer food thrice to those labouring in his fields. Every *punda* member, or his nominated family, must participate in the exercise. If either fails to do so, they have to pay a fine to the village council. The funds thus collected, are used for feasts with *satra* members.

The *Dutaru-punda* are members from one sub-clan, or willing members from different clans. The *Dhangadu-punda* constitutes young men of different *punjas* of villages, whereas *Daasi-punda* is due to young-women members. The process to utilize the help of these *pundas* are same. The *Daasi-punda* is the smallest labour cooperative intended to do small scale works. In *Daasi-punda*, young children are also invited to the annual feast, but not so in other *pundas*.

Indigenous Knowledge System

Being almost completely dependent on seasonal rain, the success of their crops, which is the family's only source of food for the year, is their greatest concern. The behaviour of rain throughout the rainy season is also variable from year to year, which has a direct bearing on the crops. Osunada (1994) studied how the Yaruba farming community of Nigeria utilizes indigenous knowledge in small scale agriculture. The farm community is rich in nomenclatures of different weather phenomena and their effects on both plants and animals. For generations, the wise have been passing on the information they have acquired in their time, through observation. This pertains to the overall performance of rain during a given year. The onset of heavy rain, withdrawal of rains before time and short and long spells of rain during the season are all related to the behaviour of insects, reptiles, amphibians, birds and mammals. Flowering and fruiting of plants and even the setting of seeds has a connection to this. The clouds, ring of clouds around the moon in monsoon sky, thunders and lightening are also some of the parameters.

The following information about rain and its connection with agriculture has been collected from group discussions with various tribal groups and villagers. It may or may not have any scientific basis. Depending on the region, however, different tribal communities interpret these indicators in their own way. They are also very careful in interpretation of existing IKS under the changing eco balance in the region. Following IKS has been recorded from tribes, I am sure there is still a long list.

The Bhumia tribes of Jeypore and Kundra blocks of Koraput have the following indicators for rains and crop performance:

- If spiders start weaving nets, rains shall stop soon. If the net is left half undone, there will be no more rain. In fact, it will start receding.
- Early in the morning if there is fog over the stream, then the rains are going to end soon.

- If there is 'jhim jhim' rain and the maggots appear in fresh cow-dung, the rains will end soon.
- If fox howls 'bubu bubu', the rains will stop.
- If the *Phurkata* or *Parli poak* insect lodges at the stem of the rice plants, the rains will continue.
- If *Bhursa kida*, common hairy caterpillars, appear, rains will happen. If they disappear rains will stop.
- When *Garad kanta* (a green cactus) flowers, and the flower goes into the 'close petal' stage, the rains will come.
- When Badali flowers bloom, rains will come.
- If the Ayabengo or Aya frog, which lives on trees, crocks, rains will come.
- If frogs enter the house, rains will come.
- If 'patpoka' appears, rains will come. The patpoka are velvet textured red insects.
- If quails weep, rains will come.
- If cobras start mating, rains will come.
- If the Dhamana snake comes out to surface, rains will come.
- If fishes in ponds, jump over the surface, rains will come.
- If cattle cough and sneeze continuously, then immediate rains will follow.
- When the small black ants are seen carrying their eggs and climbing the walls, the rains will come very soon, may be immediately.
- The *Phulaie frog*, a type of local frog when croak during night, rains are expected with in 2 to 3 days.
- After the day's grazing, if cows put mud on the head and carry grass or straw in the mouth and return home, the rains will come soon.
- If fresh cow-dung has the maggots during rainy season, and when the sun shines and flies lay eggs that hatch quickly, the rains will come soon.
- If the *Gami kida* or *Chepta kida*, an insect that is rail-like in its movements, carries mud on its head, rains will appear. If there is mud on its tail, the rains will appear late. This is predicted during rainy season, when crops are standing.
- The *Sapur poak* or *Jhilimili poak* otherwise called fire fly, when enter into house, immediate rain is forecaste.
- When the *Kejra mice*, a type of dull orange coloured small mice,

climbs trees and makes a small nest and lays litter, heavy rains will appear and continue for eight days, at least.

Plate 74: Polasha or Bodali bush (Butea parriflora)

- If it has given litter on the floor, and is seen carrying the little ones, more rains will follow.
- The *Kejra mice* are also called *Kejra Disari.* The *Disari* is the shaman. When they appear or climb trees it is a strong indication that immediate rains are to come.
- If there is a ring of clouds around the moon, and stars are clearly visible inside the ring, very heavy rains are expected immediately.
- If the radius of the ring around the moon is small, the rains shall be late. This is true of that particular time and week.
- If the rains have started on the tenth day of full moon, the rains shall be late for that season and month.
- The *Badali bush* (*palash badali*) seeding phenomena is a definite indicator of the rain sequence during the season. Most farmers believe that this a strong indicator. The *Badali bush* is found in dry upland. The pod has 3 seeds. If there are two seeds, i.e. on both ends, there will be no or less rains in the middle of the season. If the seed is at the lower end and in the middle, but not at its tip, it is indicative of no or less rains at the last stage of rainy season. The opposite is also true. This is a strong indicator.

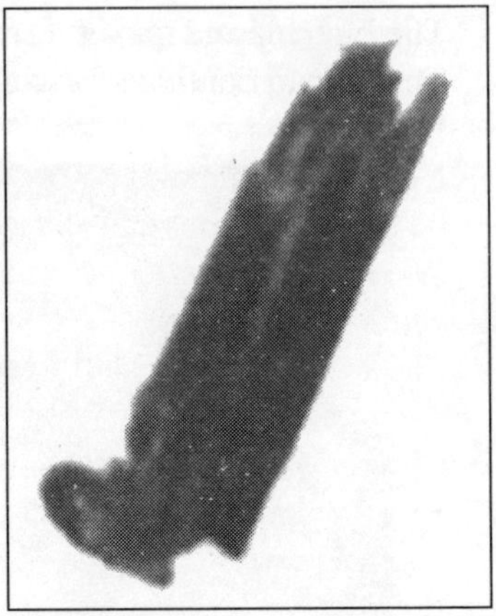

Plate 75: Darubhari poaka

- The *Darubhari* insect, also called *Katkipoak* in Malkangiri area, covers itself with a small stick like tubular structure around it. Both ends of the tube are open, with one or two, or more, spike-like structures, protruding above the open end. The size of this protruding structure above is indicative of the rain sequence. If one spike is protruding, then one great flood like rain situation for the area; and if two spikes are protruding, two great rains for the season. After completion of one heavy rain, the insect cuts off one of the spike, and after another the second round of rains, the other spike is cut-off. If there are no spikes, the rains are normal. This is a strong rain indicator.
- According to the farmers, if the cattle eats the *Darubhari* insect, the cattle bloats up and dies.
- The *Kaska Poak* insect is black in colour. It has two antennae and four long arms and legs. The insect lives in a small hole in the ground. If the hole is closed by a 'net', it suggests that there will be no rains. If it is open, rains will appear soon.
- A good rain season will follow after a very good tamarind crop.
- If Devbadni, a plant that is used to make brooms, is flowering heavily, rains will also be heavy.

Bhumias believe that the following factors affect crop performance:

- When the 'jhar'(fountains) appears in fields, the paddy will grow best.
- When mango flowers and there are clouds in the sky but no rain, the fruit will be damaged. If rain is followed by sun shine, the fruiting will be best.
- Cloudy days when *arhar* is flowering, damage good harvest.
- If the mango and tamarind flower well, crops will be good that year.

- The biri (black gram) is helped by good rains.
- The *Jalindri* in *alsi* will spoil the crop.
- If winter is very cold, the next rains and crops will be good.
- It is not beneficial if paddy grows tall and lodges, instead it should be the sugar cane that should be so.
- When the paddy is flowering and the seed is setting, a strong wind can harm the crop.
- Thunders result in bulb formation at the roots of rice plants. They are locally called, 'manga' which means barren paddy plants.
- Clouds will invite pests to plants.
- Heavy mango fruiting years, will have hail storms, which spoil paddy.

The following information has been received from Nabarangpur Umerkote regions from the Gonds, Pengo Poraja and Bhattra tribe:

- *Gaduru*, a green thorny plant, has white flowers, with large petals like lotus. The rains will follow the flowering. When the flower stocks recede (degenerate) rains will also stop.
- If the milk coloured cobra makes a sound like *ghur* intermittently, after every two minutes or so, the showers will also be intermittent in nature. But, if the sound is like *ghuree* and continues for a long time, the rains shall continue.
- If the white *kakda* (storke) sits over paddy fields, rains will stop; if it is found flying, rains will continue.
- If the *Turinja* fishes swim upstream in rivers and rivulets, rains will continue. If the fish swims down stream, however, it will stop raining.
- *Saragi* tree (*Shorea robusta*) is the common tree in the forests here. If all the leaves fall from the tree, the rains will be full and heavy for the year. But if leaves from the top and bottom of the tree fall, but not from the central branches, the rains shall be poor and discontinue.
- If the *Ayabengo* or *Aya* frog croacks, rains will come.
- When the *Jindrikida*, an insect, makes a sound like *jiee jiee*, it is an indication that rice should be sown. It is called *Khalidhan pag*.
- When the *Harada* tree flowers and fruits have attained about half an inch in size, the rice sowing period is just right.

- When a jungle bird named *Lahara chadaie*, which has white and black feathers and long tail, makes 'a metallic sound, ting ting ting', the rains will come.
- If a bird called *Dellaliti chadaie*, moves swiftly in one direction, rains follow from the opposite direction immediately
- If the new leaves on trees in the forest are dark green, the rain in the coming season shall be full and heavy.
- If jasmine flower buds appear rains will come.
- If *Kakarli,* a canine from the forest, weeps, rains will come. It makes a sounds like *Sar pauhue*. It is fond of eating village chickens.

The Matia and Koya tribe of Malkangiri believe that

- If a bird called *Kua chadaie* makes noise during the night and flies eastwards, there shall be rains; if it flies to west, there shall be heavy rains; if it flies to the north, there shall be less rains thus bunds to strengthen to retain water and if it flies south, there shall be least rains.
- When the fishes swim upstream, there shall be heavy rains.
- They also conduct a small test during Chaitra (April) month,either on full moon or dark moon days. A few dried *mahua* flower are placed over water in a leaf plate. If the flowers dip in the water, then they predict heavy rains for the year.
- *Panimariba kakada jane,*
  Pani chadiba Makada jane,
  which means the crabs know when rains will come, the spiders know when rains will stop. When the spiders spin net over the hoof prints of cattle in the paddy fields, the rains are going to stop.
- When cattle sneeze, rains will come soon.
- When Sal, Kendu and Charkoli trees sprout broad and larger than normal new leaves, it is a sign that it will rain heavily that year.
- When poultry at home lie down and come in close contact to the floor, and then it spreads the wings away from the body, rains are predicted. If one of the wing is pointing to the moon, there shall be heavy rains.
- Some predict rains on the behaviour of *jharana*. If jharana stops there shall be rains.
- There is a saying here, *malighare chelli* (goats in the house of *mali*, a vegetable farming community), *komarghare ghoda* (horses in the

house of potmakers), *tantighare mankado* (monkeys in the house of cloth-weavers), *aihtinighare keba na rahibo* (never live with such families you will be blamed).

The Laxmipur Kondhs have the following knowledge and taboos with regard to rains.

- If *Ipta* or *Kangodi*, a bird, makes a repeated sound, like *Ku*, *Ku*, during night, then there shall be immediate rains.
- If *Kedra* or *Sauria musa*, a type of mice, prepares a nest, then rain shall come soon. This is the same *Kejra* mice mentioned in reference to Bhumias.
- If *Kabri bakada*, a mammal, weeps at night, rains will come soon.
- If *Jhadi poka*, appears, rains will continue.
- If small berries, or *Chuchudel* mature, the rains will come.
- If the *Katividi*, a type of bird that lives on the banks of rivers, comes to the dung of an animal emerging from the river, the rains will stop. If their feathers change to yellow, the rains will stop.
- If *Bhabhili* (wild cat) cries, making a sound like *Bau*, *Bau*, *Bau*, rains will reduce.
- The cows calving in Bodhomas,
- The buffaloes calving in Magha,
- The hen laying in Pondo mas,
- The hen laying eggs during night are inauspicious for the family.
- The hen crowing like a cock is not a good sign for the family who owns them.

Agricultural Implements Used by the Tribal Communities in Koraput Region

Agriculture depends on implements, which are appropriately designed for the purpose. The finishing of such instruments depends on the developed skills the said tribe has acquired.

Chatodi (*Umbrella*) : It is a large headgear (hat) to save the user from rains. The normal sized one has a diameter of 84 cm and is circular. It is made up of a light bamboo framework, between which the siadi leaf is pressed in layers. It is thus light in weight. Twine strings are used for tying it firmly under the chin. This is used extensively by both men and women, specially during agricultural operations.

Tarla (*Rain coat*): It is made up of large leaves of siadi. It is 90 cm long and 80 cm wide. The thickness is about 4 cm. The green leaves of siadi are layered serially and stitched with thick siadi threads. It is used as a shield against rain and cold winds. It is also light in weight. It is worn around the shoulder and hangs down to the knees. This is used extensively by women, while transplanting paddy.

The Chatodi and Tarla are the unique and symbolic agricultural gears for all tribal farmers of Koraput district.

The tribal farmers do not use any kind of shoes. Their feet remain exposed to mud and water. Perhaps due to thorny fields and hot sand, the north Indian farmers use shoes.

Langala (*Plough*): The ploughs are simple and are largely made locally. These are made out of wood. A small iron shoe is fixed to the plough at the digging end. There are two types of plough, which are differentiated by shape and size. One type of plough is called 'Baya Nangala', and is well suited to hilly terrain and stony surface. These wooden ploughs are small and light, unlike the 'Paiko Nangala', which are large and heavy and used by farmers on the pains.

Patta (*Leveler*): Made of wooden planks and pulled by pair of bullocks, a patta is used to level fields, including wet paddy fields. Its size and weight varies, depending on the drawing strength of the bullocks.

Tangia (*The Axe*): On their way to work, or when they are headed towards the forest, the users particularly the Koya and the Kondh hang the axe on their shoulder. It is used to cut trees, chop wood for fuel at home and for splitting bamboo for thatching.

Kodki (*Spade*): Generally this is bought from the market. A wooden shaft is fixed to the blade, and is used to loosen the soil manually. It is also used for cleaning weeds, preparing canals and irrigation, or raising soil around maize plant, etc. The Desia kodki is small and designed for hill cultivation.

Illa (*Sickle*): This is a universal harvesting equipment. Its working edge may be regular or serrated.

Sabala (*Crowbar*): This is a slender cylindrical iron bar, pointed at one end. Its length varies and it used for digging. Women use smaller ones to dig tubers.

Kati (*Bill Hook*): A large knife is used to clear bushes for slash and burn cultivation. Both man and women use this implement. It is so efficient that a couple can clear a large patch of forest land in a day.

Kolki or Dhala (*Big storing basket*): These are made of bamboo. They are prepared at home or purchased from the weekly market. It is painted thick with a paste made of cow dung and mixed with soil, and dried in the shade. The inside of the basket is thoroughly smoked before use.

Kula (*Winnower*): Made of bamboo, it is used to fan and clear paddy at the threshing floor. The women use it at home to clean the grains before cooking.

Ghurna (*Gushing Wheel or Grinding Wheel or Grinder*): Made of wood or stone, it is a hand-driven grinding equipment to make flour at home. It is made up of a pair of circular heavy stone/wooden blocks. While one is fixed, the other is mobile and fixed around a wooden pin. The wooden grinder is common with tribes depending on millet as staple food. The stone ones are normally used to grind grain to flour.

Musulo (*Pestle*): It is made of wood, with a small iron ring at its end to de-husk paddy.

Most villages have an ironsmith and almost all tribals are carpenters,

Plate 76: Tribal women hand pounding paddy

Plate 77: Hand stone grinding of ragi

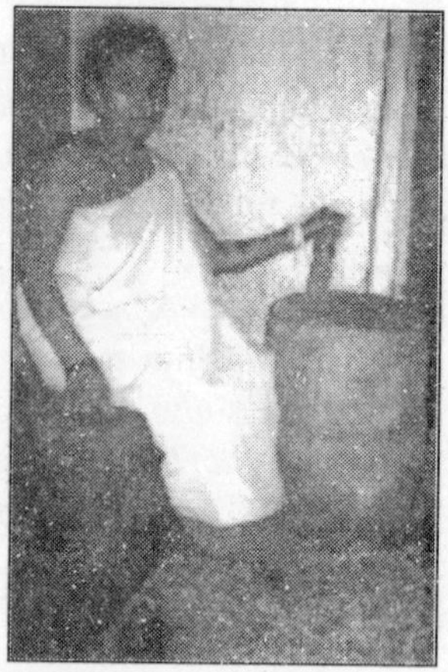

Plate 78: Wooden hand grinder

Plate 79: Drying vegetable over heath, Paranga

Plate 80: Baya langala of Koraput

Plate 81: Juadi and langala, Kotpad

Plate 82: Tribal women's rain dress talari and chatodi

Plate 83: Large bamboo bin for preserving paddy

who prepare their own implements during the season. All above implements, except for ploughs and levelers, are sold seasonally in weekly markets.

Tribal Women at Home and in Agriculture

The women of every tribe in Koraput district are the pivot in the family and, therefore, to the community. This is due to their role in socio-culture and socio-economic frame of the community. Contrary to other neighbouring societies, on her marriage, a tribal woman receives a bride price from the groom's family. This may be in the form of rice, wine, pendum, vegetable and livestock. The bride gift is thus made into a good feast. This itself is indicative of the respect and importance that a woman gets in the tribal social system. They hold a higher status than men and have the right to divorce as well as a say in their marriage. They take equal part in the social and economical decision processes. They have a very high degree of freedom within the household and social boundary. They enjoy a higher degree of support from women in situation under threat from her husband or others.

Patnaik (1989) has reviewed the position of women in Kondh society. He wrote, 'Now as a housewife she is able to on account of her services, control her husband. She joins the men at the Panchayat.' Routray (1987) reported the status of women among the Dongria Kondh of Koraput district. The author concluded that, Dongria Kondh women are debarred from certain religious and managerial functions, their social status is very high. It is also established that in various aspects of social life women enjoy status and freedom, such as in the choice of a husband, pre-marital sexual licence, seeking of divorce and so on. But in other few aspects, such as in political sphere, they get low status.

On the other hand, Singh (1993) remarked in general terms that the tribal women's status is characterized by over-work, illiteracy, sub-human living conditions, high fertility, high malnutrition and absence of modern human health care.

Most tribal women of Koraput are forbidden to hold the plough to cultivate the land, or sow seeds for field crops; however, the Bondo, Koya, Didayi and most Gond tribal group women can perform these tasks.

Owing to the changes in the agro-economic scenario, the tribal women and her role are fast witnessing a change. The forest base wherever available, is facing fast depletion, affecting the food gathering activities as well as the income through sale of minor forest produce. At the same time, however, in some areas opportunities are appearing from new industries and development projects.

Menon (1987) investigated the impact of deforestation on women in tribal areas of Orissa. He recommended measures towards reforestation. The poor tribal women within poor households are particularly dependent on common property resources (CPRs) for their livelihood. But these CPRs are being lost at a rapid rate in the tribal areas due to population pressure (Roy and Tisdell, 1993). Jain (1992) studied the pattern of inter-gender involvement in animal husbandry operations in the Hissar and Kurukshetra districts of Haryana. Bhati and Singh (1987) had estimated the total labour inputs of household and found that the women's labour accounts for 61 per cent of total farm work.

There is scanty information available on the aspect of tribal women and their role in agriculture, household activities and economic pursuits. Mohanty (1991) has given the calendar of economic activities for Kuttia Kondhs of Kalahandi district, living in Thuamul-Rampur block. Mishra (1991) stated that Nishi women of Arunachal are the axis around which the Nishi economic life revolved, yet their role in the social and political spheres was almost negligible. Almost the whole burden of slash and burn mode of agriculture was on Nishi tribal women.

The economic well-being of tribal families is mostly dependent on women, who contribute a lot; but under the changing situation it is not possible for them to take additional responsibilities. The tribal women in Koraput and Bastar districts work 12 hours a day.

In Rajasthan, tribal farm women work 15-18 hours a day. They work longer hours than men. There was a significant difference in home and farm activities between tribal and non-tribal women (Rathore and Gaur, 1996).

The children are loved and cared at home. The family offers several promises, offerings and prayers to various deities, for the well-being and good health of children.

From the age of seven, girls start working at home and assist in several ways, according to their capacity.

Childhood

In the absence of parents who are occupied in agricultural activities, they carry and care for children at home. Generally elderly persons in the village assist them.

- Weather permitting, children usually play for few hours every morning and evening. They even participate in every village dance.
- With elders, they collect drinking water from spring or tube well in small vessels.
- When they are eight years old, girls accompany women to the market and forest. The arduous jobs of agriculture are carried out after they are at least 11-12 years old.
- They learn to cook and get tips and advice on their future role from their mothers and grand mothers.

Food Collection from Forests and Food Production Skills

- The girls learn group dynamics and food gathering from the forest. They learn and collect all kinds of vegetables, fruits, berries and eatables. They learn to identify poisonous plants. They follow the rules of jungle walking, climbing hills, and safety from poisonous snakes, tiger, leopard and bear.
- Before their marriage, girls learn enough about folklores and literature pertaining to the forest and the environment.
- She knows taboos for men and women in the society

Podu Land Cultivation Schedules (where Applicable)

The family identifies the area, while the village committee grants the permission. This is the most arduous job tribal women carry out. Women guide the families in implementation of total sequence of podu cultivation including the ceremonies. Women clear the bushes and light the fire. They clean out the weeds, make the bunds and clean the channel to allow the rainwater to flow freely. They also take out food to the men posted on crop-water duty. Women start the first harvest of the year by hand harvesting the first small millets with a knife. The harvesting process marked by a brief ceremony and offerings. While women join threshing of all the crops, men only do rice threshing. Women are also responsible for the storage of the grain. Women are well acquainted with processing the food material collected from the forest, like bamboo shoot, tuber or kondas, flower and fruits of *mahua*, Kendu and many others. Women are also knowledgeable about common medicinal plants in their environment and their use.

Backyard Garden

It is the women's responsibility to sow seeds in time, to look after creepers and provide them support to climb, and to protect them from livestock, poultry and nocturnal birds, like bats, etc.

They harvest beans and little gourds almost every alternate day. They are in charge of processing of various vegetables and cooking.

The collection and transport of surplus vegetables to weekly markets for sale to earn cash in post-rainy season, is also a woman's responsibility.

Kharif Cultivation

The agricultural tasks entrusted specifically to men are not mentioned here, only those that are assigned to women, like:

- Weeding of agricultural plots.
- Transplanting of paddy saplings, which takes between 15 to 30 days. Where paddy land is available, the women of the village usually labour collectively.
- Weeding plots and transplanting paddy. This involves standing in mud and water for up to seven hours a day, on an average, and bending at a 30° angle at the waist.
- Harvesting and carrying crops to the threshing floor.

Early Rabi

- Transplantation of ragi (in Laxmipur & Rayagada area).
- Harvest of *alsi*, *kandula*.
- Threshing ragi.

Rabi Season

- Weeding of oil crops.
- Weeding of maize, where irrigation facility is available.

- Pot irrigation of tomato, brinjal, tobaccos on alternate days mostly near river beds and in the backyard.
- Collection of tomato and brinjal for food and sale.

Early Summer

From early morning, women go out to collect *mahua* flowers from the forest and around the villages (end of February to April).

Social Functions

Marriage ceremonies: It is a woman's duty to feed friends and relatives who come home on such occasions. They prepare and carry food and drinks to the guests. They prepare *Pendum* for the family, to use at least once weekly during summer.

Late Summer

FUEL WOOD COLLECTION

- Collection of fire wood and stacking. Going for wage labour work mostly to development projects run by the government.
- They can be seen laying railway tracks and roads, working in the forest department and in soil conservation plantation work, etc. Some women work at stone quarries, break stones to small chips and selling to contractors.
- Many women also collect, process and pack Kendu leaves for sale.

Weekly Market

The women attend weekly markets that are held close to the village. They visit once a month, for eight months of the year. And during post harvest season, they visit weekly markets each time they are held. Women carry head loads or hand loads only.

Social Functions

RELIGIOUS

- They use a calendar to keep track of various religious events, informing the man in advance about the requirements for each occasion.
- Villagers hold joint annual or triennial celebrations for prosperous

agriculture, and good health of the community. Offerings are made to the dead, and other valuable resources like forest, hillocks, village boundary, springs, and rains, etc., are worshipped.

- Women collect coloured soils from different places, and treat them with barks of trees and then paint the walls and floors of the home before important community celebrations.

ANNUAL THATCHING OF HOUSES

- Cut and transport head loads of dry forest grasses to home.
- After man have finished thatching, they clean the house and dispose spoiled straw and roof materials.

NEW HOUSE

- Women take equal responsibility in preparing mud and making walls and floor during the construction of a new house.

Animal Husbandry Activities and Roles of Women

Housing	Early in the morning, they clean the cattle byre, the pig sty, and the sheep and goat houses. Between 5.30 and 6 a.m., let loose pigs and poultry.Between 7 a.m. and 9 a.m., they let loose cattle and buffaloes, sheep and goat and hand them over to graziers. Taking stock of the livestock when it returns in the evening and reporting losses, if any, to the men.
Poultry rearing	They are responsible for the pre- and post-hatching duties, ensuring the chicks are protected from predators, heavy rains, sun and cold winter nights. They must also protect their backyard crops from their poultry and pigs. They have to ensure that their pigs do not run into others' backyards or homes.
Animal breeding	The breeding knowledge is received through men, from the graziers.
Management	Feeding of pigs and chicken, where available, at least once or twice a day.
Marketing	Take only poultry for sale to the market; sale of

	other livestock is the men's responsibility. The tribal women go to weekly market for business and sell vegetables and other agricultural produce. About 80 per cent of the sales persons in a weekly market are women.
Health	Feeding newly born lambs, kids, does and ewes, and report to men if the animals are affected by any disease, sickness or wound.
Family seed collection & preservation	Seed collection, processing and preservation for the next agricultural season is a very serious job performed by women.
Decision-making process	The decision-making process is a collective one, in which both the husband and wife, and indeed the entire family, takes part.
Daily routines at home	House maintenance cleaning of cattle byre, sheep and goat sheds cooking and serving utensil washing filling drinking water and carrying it home fuel collection (may be seasonal or weekly) cooking twice a day packing food for those on watch and ward duty in the field or in the forest clearance at noon bathing children washing clothes feeding children and taking care of them care of the elderly care of the sick and pregnant women care of one's husband
Food processing	Processing mango kernel Processing of ragi to flour. Processing of paddy to rice Cleaning of tamarind seed for pulp Processing *mahua* flowers and fruits Processing of root crops, bamboo roots to detoxify

Cleaning lentils
Collection of dry lentil plants for fuel at home
Preparing beer for the summer, from fermented rice and ragi
Processing seasonal food for preservation
Processing of turmeric bulbs with men.

As per FAO (1998), women produce more than 50 per cent of humanity's food supply globally. In sub-Saharan Africa, they provide up to 80 per cent of staple foods. In Asia, they perform up to 90 per cent of the work in rice fields. The women in tribal world produce and provide up to 80 per cent of the food supply.

The tribal women have communicated that operations in slash and burn cultivation, the weeding in crop culture, husking and grinding of grains, and collection and transportation of firewood from the forest and hills are most laborious and time consuming jobs. The collection and transportation of drinking water in the deserts by women is an equally difficult job. This needs serious thinking.

Tribal Community Ethics and the Modern World

There are social moral and community ethics, which are in built among tribal societies. They have been exploited by others and need to be looked afresh under the changing world situation. These are:

Rano

This is an 'oath' or swear in the name of god before any formal statement by tribals.

Although the tribals stick to their word in most cases, others take advantage of the system. When a business deal is struck regarding the sale of farm produce, between the tribals and a businessman, who may be an outsider, the 'rano' is raised with the advance of a few rupees. After this, prices cannot be changed under any circumstances, not even if the villagers discover that they have been cheated and are getting a lower price.

The 'rano' is in the name of god or family members. The tribals fear that non-adherence to 'rano' may lead to misfortune at home upon the village.

Jati Jibo or Outcasting

If any tribal person is beaten by a person belonging to a scheduled caste, or accepts food or water from such a person, or is bitten (even by a dog), marries outside the tribe and into a scheduled caste, gets ridden with maggots, is beaten by shoes, goes to jail or is remanded to police custody, is found in an intimate position with someone of another caste, such a person is liable to be declared an outcaste by the community. Every member of the tribe, therefore, is aware of the various restrictions.

When a person is declared an outcast, his/her entire family is also treated thus. They are not allowed to participate in community matters after that.

Except in cases of marriage into a Scheduled Caste (SC), a declared outcast and his/her family can be accepted back into the community. In some tribes, there is a hereditary position called *Bhato Nayak, bhato* meaning cooked rice. This person is in charge of several villages of the tribe, and performs a special function to admit declared outcasts, and their families, back into the fold. The outcaste has to pay a fine, have the head shaved and organize a feast for all the members of the clan and the village.

The cost of organizing such a feast is extremely high. As a result, people are generally careful not to flout the rules.

Zamindars also saw this as a handy tool, with which to recover their dues. They used to send a SC person to the defaulter's house or village. The fear of being declared outcasts haunted the tribals, making them easy preys for the moneylenders.

It is high time that the tribal leaders are educated about the issue and convinced to stop its practice.

Tapu

This is also a kind of outcasting phenomena seen in Koya society, which means 'social do-nots'. These rules apply more to women than they do to men. For instance, women cannot accept food or water from a person of another tribe or caste, no matter how high that caste may be. The use of narcotics like opium, *Canabis indica* (ganja) comes under this sanction among Koyas.

Sometimes 'Tapu' is enforced in tribal society by family and the community.

Thefts

From the time of the British, this has remained one of the problem areas in the Koraput district. The theft of cattle, specially in the Nabarangpur region, is well known and natives have made many representations to the administration on the matter. As most such thefts were committed for the hide, most cattle owners started inflicting scares on the bodies of the animals, to render their skins useless for thieves. The practice has since been discontinued.

Among the Koyas, theft was practically unheard of till the 1980s and people usually left their doors knocked. Theft of salt, however, was known.

Another recent problem has been the theft of electrical wires in the rural areas of Koraput. This interferes with the supply of electricity and the activities of the rural programmes for agricultural development.

Diesel pumps, which are used for irrigation, etc., are also under threat and a common target of thieves.

Since modern agriculture is energy-related, theft of electrical wires and power transformers should be a matter of serious concern and steps should be taken to discourage such practices. It is important to create awareness about the issue among the people.

Exploitation of Agricultural Produce

According to Patnaik and Das Patnaik (1982) Schedule Castes exploited the Dongria Kondh agriculture economy. The Schedule Caste population in Koraput are mainly involved in non-agricultural pursuits which are inadequate for living, so they try every way possible to make living from the neighbour tribes.

They act as middlemen in livestock markets between tribal farmers and others.

They also give loans to tribals mostly at 10 per cent interest per month.

The tribals cannot sell skin and bones of animals to anyone directly except to Doms. Defaulters will be declared outcasts.

The tribals are quite aware of the situation and understand that they are being exploited. Yet, they cannot avoid it.

Farming Systems and Koraput District

Ruthenberg (1980) has defined farming systems as under the described properties of farms applied as each unit, in each case there is a system which transforms inputs into agricultural outputs and which undergoes changes overtime. He has further described seven farming systems in tropical climate. They are Shifting cultivation systems, Fallow farming systems, Ley systems, Dairy farming systems, Permanent Upland farming systems, Arable Irrigation farming systems, Perennial crop farming system and grazing system.

However, MacArthur (1980) has described the tropics as a high potential from photosynthesis advantage, on the other hand, it has constraints due to the natural environment as climatic and biotic difficulties.

FAO, South Asia Farming Systems and Koraput

The farming systems have been classified on the basis of criteria resource base, dominant livelihood (main staple and cash income source and the balance between crops, livestock, fishing, forestry and off-farm activities), and the degree of crop-livestock integration. And the study was undertaken by Watherhogg et al. (2001) for South Asia under FAO. They have shown eleven broad farming systems and were of the opinion that there is a considerable number of diverse, less important farm systems which do occur within each broad farming system. Even the systems overlap each other in some instances. The broad farming systems according to them are rice farming system, rice-wheat farming system, rainfed mixed farming system, dry rainfed farming system, highland mixed farming system, pastoral farming system, tree crop farming system, urban-based farming system, costal artisanal farming system and sparse farming system.

However, of the South Asia farming systems, rice farming system, rice-wheat farming system, rainfed mixed farming system and highland mixed farming system are seen with great potential for agricultural growth and poverty reduction in the coming three decades.

The Koraput district can be classified in the light of the Watherhogg et al. (loc. cit) classification as rice farming system, highland mixed farming system and rainfed mixed farming system.

The wet rice growing regions of Koraput plains, like Jeypore, Kotpad, Nabarangpur, Gunupur and Malkangiri, can be classified as rice farming system.

The hills of Koraput district are classified under highland mixed farming system, whereas rest of the area, which is not covered by the above two farming systems, is termed as rainfed mixed farming system. The forest-based farming is one slash and burn cultivation system that can even be classified within highland mixed farming system.

Rice Farming System of Koraput

Rice is invariable grown in the wet season. Rice, or another less water demanding crop (e.g. coarse grains, oilseeds, legumes, vegetables), is also grown in the dry season.

Although some farmers are aware of the benefits of fertilizers and improved seeds, these have not been accepted enthusiastically elsewhere.

The rice farming system has considerable in built diversity. Aquaculture, cattle and poultry supplement the system.

Highland Mixed Farming System of Koraput

The predominant crop are millets/sorghum in the hilly fields, maize and beans in the backyard at home, legumes (pigeonpea and dongarani), and the oil crop is predominately the *alsi*. The rice is grown upland as well as in wet lands. There are special situations on hills, where horticulture is an essential component for sustainable food and resource generation. This is specially seen in Dongria Kondh and to some extent in Hill Sabaras.

Watherhogg et al. (2001) wrote, 'The system (high land mixed farming) intrinsically less able than the low land areas to provide production responses in food crops, but they are likely to see a measure of diversification into high-value produce as market access spreads. In the meantime, the seneri poverty present in these areas will probably lead to continuous seasonal and permanent migration, especially of young people.'

The farmers in highland mixed farming systems utilize the least fertilizers, and the high transport costs easily exceed the material cost. The extension system of agriculture is poor in these regions.

Watherhogg et al. (loc. cit.) were of the opinion that these areas are occupied by tribal societies who originally practised shifting cultivation. The most important issue here is increased population pressure, both naturally and through immigration by 'low landers', which can make the already fragile system unsustainable. However, there is no immigration from lowlands here, although that of the high landers is apparent in Koraput district.

These areas have experienced little development assistance, and the trend has been the general deterioration in incomes, living standards and the overall environment, as population has increased. The reduction of forest cover on the upper slopes, unimproved and unchanged agricultural system, and continuous cultivation on steep slopes, which now have thin or poor soils have led to severe poverty. Incomes are very low. Women have to walk long distances, and up the hills, to collect firewood.

Rain Fed Mixed Farming System

Watherhogg et al. (loc. cit.) wrote,

> A large proportion of the rural population in the system lives in poverty. The agriculture is oriented towards subsistence; while most areas are poorly served by infrastructure and services and are remote for markets. Agricultural extension services in these areas are typically weak, farmers mostly use traditional technology with a strong bias towards risk avoidance. Land tenure is often an issue and farmers may not have sufficiently clear titles to their land to be able to use it as collateral for obtaining institutional credit.

This also holds good for the regions of Koraput district.

Agro-climatic Zones and Farming Systems of Koraput as per ICAR

The state of Orissa has been classified in ten agro-climatic zones. Koraput comes under the Eastern Ghats Highland Zone, South-Eastern Ghat Zone, North-Eastern Ghat Zone and Western Undulating Zone.

This classification has been done by the Indian Council of Agriculture Research.

1. EASTERN GHAT HIGHLAND ZONE

Identification of agro-ecological situations (farming situations) of the zone is mainly due to altitude. The three farming situations of the zone are indicated in the Table 4, along with the blocks covered under each situation as well as the major crops.

TABLE 4: FARMING SITUATIONS AND BLOCKS COVERED UNDER EACH SITUATION OF EASTERN GHAT HIGH LAND ZONE

Name of the farming situation	Blocks covered	Crops
Farming situation I (Elevation 600-1000 m above MSL)	Koraput, Dasmantpur, Semiliguda, Nandapur, Pottangi, Laxmipur Lamptaput	Ragi, rice, *alsi*, small millets, *arhar*, sweet potato, dangarani vegetables
Farming situation II (Elevation 300-600 m above MSL)	Narayanapatna, Bandhugaon	As above + *kulthi*, cotton, tobacco, castor, chilli, vegetables
Farming situation III (Elevation 150-300 m above MSL)	Umerkote, Jharigaon, Chandahandi, Raighar, Nowrangpur, Nanda-handi, Tentulikhunti, Papadahandi, Kosagu-muda, Kotpad	Same as I, except *arhar*, sweet potato, dongrani, *kulthi*, sugar cane, sunflower and vegetables

The climate is warm and humid, maximum (mean) temperature in summer being 34.1° C and mean winter temperature is 7.5° C. About 10 to 20 cm of rainfall during October helps the second crop.

2. SOUTH-EASTERN GHAT ZONE

Four farming situations are delineated based mainly on elevation, rainfall, soil types and irrigation potential. The climate is warm, with mean maximum and minimum temperatures being 34.1° C and 13.2° C respectively.

A brief description of each farming situation, and the revenue blocks covered under each situation, is presented in Table 5.

TABLE 5: FARMING SITUATIONS OF SOUTH-EASTERN GHAT ZONE AND THE BLOCKS COVERED UNDER EACH SITUATION

Name of the farming situation	Blocks covered	Crops
Farming situation I		
Medium rainfall (1222 mm) high elevation	Jeypore, Borigumma, Kundra, Boipariguda	Rice, ragi, *alsi*, *kulthi*, maize and vegetables
Farming situation II		
Medium rainfall (1247 mm)	Mathili, Khairput, Kudumulgumma low elevation	Rice, ragi, small millet, maize, *alsi*, *moong* & vegetables
Farming situation III		
High rainfall (1328 mm) low elevation	Malkangiri, Korkunda	Rice, ragi, *til*, groundnut, *moong*, maize, *mestha*
Farming situation IV		
Low rainfall (954 mm) low elevation	Kalimela, Podia	-do-

3. NORTH-EASTERN GHAT ZONE

The Rayagada area comes under the North-Eastern Ghat Zone. The climate is hot and moist and sub-humid with mean maximum temperature of 37° C in summer and mean minimum temperature of 10.4° in winter.

4. WESTERN UNDULATING ZONE

The Dabugaon block of Nabarangpur comes under this zone. It is classified as having red and yellow soil, high rainfall and medium elevation. The rainfall is 1200-1600 mm. The climate is hot, moist and sub-humid type. The crops are rice, ragi, black gram, *til* and vegetables.

TABLE 6: FARMING SITUATIONS OF NORTH-EASTERN GHAT ZONE AND BLOCKS COVERED UNDER EACH-SITUATION

Name of farming situation	Blocks	Crops covered
Farming situation III Red loam soil, moderate rainfall, high elevation (500-1000 m), rainfed	Gudari, Muniguda Chandrapur, Bisam-cuttack, Kalyan Singpur	Rice, ragi, small millets, jowar, maize, *arhar*, dongarani, black gram, niger, *alsi*, *til*, custer bean, mustard, turmeric and vegetables
Farming situation IV Red loam soil, low rainfall, moderate elevation (300-500 m) moderate irrigation	Ramanaguda,Gunupur Padampur, Raigarh, Kashipur, Rayagada Kolnara, Guma	As above

Crop-Livestock Farming Systems of Tribals

The crop-livestock farming system described here is also based on different kinds of food production systems, which are:

- Backyard farming system
- Life support crop farming system
- Scarcity period food system
- Famine saviour plants in the system
- Slash and burn cultivation system (*Podu*)
- Mixed species animal husbandry system

Backyard farming, which is the general rule, is managed entirely by women. The housing pattern and terrain dictate the size and location of the backyard farm land per family. The hill top dwellers have a special patch of land close to the village, which serves the purpose. Generally, it is behind the house. It is a vital system to produce various vegetable and a small amount of crops for cash. The Kondh are most skilled in utilizing various plant species in their small backyard garden.

The Life support crop farming system consists mostly of small millets and legumes of different durations. These support the family during the period of annual food scarcity. Thereafter, different food crops appear in continuous sequence, till the end of the last harvest for the season. The food production cycle starts from mid-July and lasts up to November-December, and in some cases up to March.

Scarcity period food system is most vital for survival. Periodically, hill dwellers face shortage of food at home. The degree of severity varies, and it depends on many factors. But mostly non-conventional food resources of the forest, such as bamboo shoots, different tubers, berries, mango kernels, and wild fruits, etc., relieve the situation.

The famine saviour plants in the system of tribal life, for all times, are wild mango, jackfruit, tamarind, *mahua* fruits and flowers and many others. Several wild fruits, flowers, tubers, leaves, seeds and shoots come under this system. They are available as food to the families during the time of acute food shortage. The cash obtained from the sale of such plant produce helps the family food requirement.

A thin line separates the scarcity period food system and the famine saviour plant system.

The slash and burn cultivation system (Podu) is mostly followed by tribes living on hills and in the forest. This is the most depleted system when soil fertility is lost. The Dongria Kondhs have a horticulture base added to podu cultivation, which is very productive. The Sabara tribe has a well developed, meticulously planned, multi-plant species farming within the slash and burn cultivation system and is sustainable source of food supply to families.

The mixed species animal husbandry system is the rule, where few animal from different species, are reared by tribal families. It is mostly a scavenging system base.

Backyard Farming System among Tribal Groups

The backyard farm is also known as compound farm, garden farm and household farm. All tribal families love to have at least a small backyard behind their homes. The women take care of the plantation and management and men make and maintain the fence around it. Hahn et al. (1990) examined the compound farm in African agriculture. They opined that it is an indigenous farming system, which can be further improved with research, leading to improved sustainability. Hahn (1991) was of the opinion that it is an unexploited and untapped potential for family food security in Bangladesh, Thailand and Africa. He stressed that, using the knowledge that the farm women had acquired, the system could be improved to provide additional income, regular food supply through the year and improved diet for the respective families.

The backyard farming system can be classified for tribal farmers as:

- Open manure collection pits with pig and poultry production base in the backyard.
- Temporary enclosure—Maize and creeper beans production for *kharif* season only.
- Permanent enclosures:
    - Mud compound wall, fruit tree and seasonal creeper vegetable base (pumpkin, gourds, beans, etc.).
    - Economic tree like bamboo and other fruit tree on the fence and vegetable base backyard.

Open Manure Collection Pits with Pig and Poultry Production Base System

Topographically, the Kondh tribal village has its unique utilization of the backyard farming system, namely the pig and poultry feed resource

Plate 84: Kondh backyard with open manure pits—pre-monsoon

Plate 85: Kondh backyard with open manure pits—winter

and manure production base system. The cattle and pigs are housed behind the houses, which also run in a close wall system, allowing no space to the livestock to come to the main street of the village. Each house has its proportional space behind the house, where they collect the cattle, buffalo and pig manure daily, after cleaning the byres and sty. These are well demarcated, and shallow pits. The pigs and poultry scavenge the whole day in these pits for insects, maggots and other unicellular materials that are produced here. The inhabitants clear the manure pits periodically before agricultural operation every season, and start laying the dung again, layer by layer, along with the daily refuse from the kitchen.

It is an air layering system of producing organic manure that, together with the pigs and poultry production, represents the traditional farming system of Kondh. Several studies have shown that animal manure is widely used to maintain the fertility of cultivated fields in the savannah region, south of the Sahara and elsewhere.

The productivity of village manure pits is dependent on holding size, and seasons of the year. It has direct bearing on pig and poultry production through supply of feed resources; and it contributes organic manure to the agriculture production system. Though the primary aim of the farmer here is manure, but the system has well assimilated the pig and poultry production in it.

This system as it is sustainable, considered from food production support point of view. The option to upgrade existing manure pit production system through 'anaerobic compost preparation system' in the backyard of Kondh villages should call for serious thinking. The spray of insecticides over these backyards will have an adverse effect, at least on the available feed resource for pigs and poultry.

Research on growth of flies, mosquitoes, white ants and other insects under such environment, and how villagers are adopting it, are important and can only be confirmed by studies in the villages.

I think from my observations that the village poultry and pigs remain all through the day in these pits and always in search of insect food, thus there is very little chance that insects harmful to human being can escape from them, provided the animals perish to a significant number due to epidemics.

The closed houses of the Kondhs are like long, dark tunnels, with one door in the front and one at the back. The house has no windows,

Plate 86: Inter-crop of maize beans

Plate 87: Tunduri creeper in Jeypore plains

Plate 88: Semi beans in Kondh backyard

and the smoke from the kitchen allows neither flies nor mosquitos to breed inside. Studies conducted in Africa, about the connection between ruminants and malaria, suggest that mosquitos like to feed on livestock. Instances of malaria are inversely proportional to the size of livestock in a given area. Therefore, keeping these animals near the manure pit in the backyard, is one way of keeping mosquitos away from the house.

Mishra and Ramakrishnan (1982) remarked, 'the compost pit (Khasi village) is a very important component in the functioning of the village eco-system as it provides feed for secondary production and generates organic manure both of plant and animal origin which is cycled back into agriculture'.

Animal husbandry formed an important link in the food chain through utilising the garbage and vegetable waste of the agriculture system. The forest, apart from providing the basis for agriculture, also meets part of the fuel requirement of the village.

Temporary Enclosure Backyard System

This is mostly the situation among almost all tribes, depending on rainfed farming, who are also usually poor and resourceless. They make small enclosures around their houses and grow principally maize and creeper beans or other creeper *kharif* crops. The fencing is only to prevent the entry of pigs, sheep, goat and poultry, and even wild bears, into the enclosure.

Permanent Enclosures Backyard System

In villages around Nabarangpur, Jeypore, and Kundra blocks, fencing around the houses is necessary. The front portion is almost a mud wall and the backyard is enclosed with bamboo, and fruit trees like mango, tamarind or jackfruit between thorny bushes. This also serves as a fence. This way they have bamboos for house thatching and they get fruits from the trees. They take care of the plantation in such a way that the shades do not interfere with the growth of vegetable crops from the space available.

The well-to-do farmers of Omnatyas, Bhumias, Bhattras, and Halbas prefer mud walls around the house and the backyard. They plant fruit trees, bamboo, and *kharif* creepers inside. The cattle, sheep, and goat are housed in front of the house,and are not provided any access to the backyard. There are no pigs in this system of backyard farming.

The beans are a very vital food resource for the tribal families. There are several varieties of beans. They have different colour, fluorescence, shape and size and skin colours of pods, and even the shape and size of seeds varies. In some of the beans the skin of the pod is fleshy and tender, even juicy. They are cooked along with the seeds as curry, particularly the creeper 'Bailo type', which has flat pod and seed. This is the common type of bean. The family collects some pods from the backyard on alternate days and cook them up to the winter. Some families maintain the creepers as an annual plant.

These beans are grown close to the house, in the backyard. These are sown during June, and the creepers are supported by 'Ranjas', a small indigenous bush, cut separately for each creeper, from the neighbouring forests. Until they have grown to about 3 to 4 ft high, they are protected from poultry, sheep and goats. Nowadays, these beans are also inter-cultivated with maize. They grow over maize plants, and once the maize cobs are harvested, the stems serve the bean creepers.

A study on bio-diversity of creeper beans in Koraput district will be highly interesting, although none has been conducted yet. Maize is one essential crop with all tribal farmers who have some land behind the house.

In the plain land villages of Jeypore, Nabarangpur, Mathili and Kundra areas and other places as well, the families grow one or two vines of 'tundri', either in the backyard or in the frontyard. This vine trails on a 6 x 4ft *pandal* called *chamunda*, which stands on a four-pillar support. On alternate days, the families harvest the tundri fruits, sufficient for vegetable requirement of the family during the summer and monsoon. They have a system to wash utensils under the pendal, so that the vine gets regular water. This is a functional system and every tribal home in the said area have adopted it.

Nowadays, before the *kharif* cultivation in some villages, even the pig-loving tribes prefer to sell their pigs and safeguard their crops in the backyard. It also helps people avoid feuds with their neighbours. After the harvest is over, they buy pigs again and rear them for the rest of the year. In some villages, people keep pigs in enclosures made around the stys, and supply little feed to them. Before the *kharif* cultivation, families keep only a few hens and roosters for breeding,

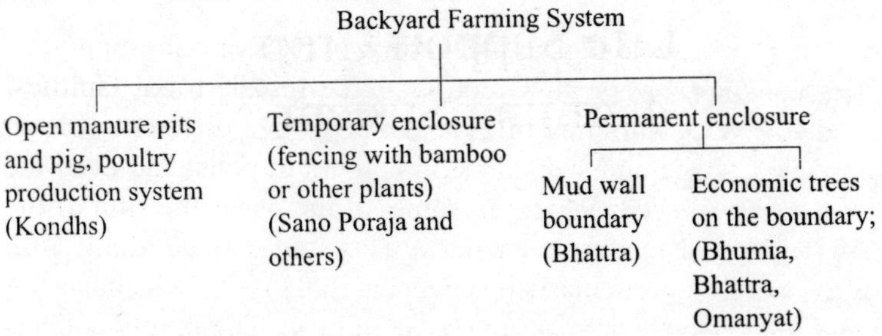

and dispose the rest. Some families even tie them temporarily around the house, in order to stop them from damaging germinating seeds in the backyard.

Life Support Crop Farming System

The tribals cultivate a few specific crops that have varying periods of maturity, which suit their farming and socio-economic environment. These valuable crops, usually pseudocereals and millets, are rich in minerals and vitamins. Pulses are rich in proteins, specially some aminoacids. Vegetables like pumpkins and cucumbers also fall in this category and help the Koya, Didayi, Sabara and other tribes to tide over the early agricultural seasons. These crops are termed life support crops, particularly the small millets, because they provide food to the family during times of acute shortage, when there is no other available food. These species are under utilized plants.

The small millets are finger millet (ragi), *Eleusine coracana*, kodo millet (kodo), *Paspalum scrobiculatum*, fox millet (kangu) *Setaria italica*, little millet (kutki), *Panicum miliare,* and barnyard millet (suana), *Echinochloa frumentacea* and bulrush millet (*Holcus spicatus*). Except finger millet, all above millets come under the *Paniceae* family, whereas finger millets come under *Chlorideae.*

The legume plants of significance for them are *arhar* (*kandula*) *Cajanus indicum*, rice bean or kutting (dongrani), *Vigna umbellata*, and horse gram (*kultha*) *Macroty loma uniflorum*.

Compared to other small millets, both suana and panicum are the dominant types. Grown on the hills, the suana grow very fast and have a strong root system. This is a great help during heavy rains on the steep slopes. It is difficult to find alternate crops in such steep gradient conditions. This is what happens in Eastern Ghat hills particularly in Kashipur area of Rayagada.

The earliest varieties of small millets are ready to harvest from mid-July, but most of it is harvested on hills on 15 August. The harvest is the first in the new food cycle of the year. This gradually builds up with larger quantity and different species of plants, like early maize,

Plate 89: Pencium millare/Kakigumma

Plate 90: Sorghum/Sabara hills

Plate 91: Suana long duration (violet head), Laxmipur

early dashara ragi, dashara upland rice, followed by medium duration ragi and other millets and ends up with long duration varieties, including rice, *kandula*, jowar, etc.

The *kandula* is the long duration type. The *kandula* is the support crop, which saves the family in case millet and other cereal crops fail. It has adapted well to the area and conditions. The *kandula* adds up to 40 kg of Nitrogen per hectare to the soil.

Rice bean is mainly confined to the tribal regions of north-eastern hills and western and eastern ghats in peninsular India. In the north-eastern region, rice bean is grown predominantly under rainfed conditions, in a mixed farming system, under shifting cultivation, or kitchen gardens and backyards. The Koraput hill tribes grow rice bean, which grows profusely without any rhizobium culture, even under adverse soil fertility status. In hills, good biomass production for green fodder can be obtained from rice bean during September to December. The original cultivars are creepers type, and improved selection from research are mainly higher grain yielding and non-creeper type.

Being the short duration type, the local maize varieties of the Didayi tribe of the Malkangiri are harvested as the first family food of the season. The Kondhs of Rayagada grow local Pearl millets (Ghantia) *Pennisetum glaucoma*, which is also a short duration crop. This helps the family at times of acute food shortage. It is grown under mixed stand.

The creeper beans form the next vital crop. These are simbo, jhota and bailo, which are locally grown. They are available as food to the families after October-November only.

The tribals do not follow the general cropping patterns of the plains of Orissa, where biri and *moong* follow rice. However biri is grown as a mixed crop in Bonda hills, as also in Malkangiri. The *moong* is also preferred in many areas of Malkangiri.

Alsi (Niger) is an important oil seed crop and extensively grown by tribals of Koraput. It amounts to 25 per cent of the total *alsi* produced in India. It is cultivated on hill tops and on hill slopes, particularly on well-drained loamy soils. It performs under poor soil conditions, is resistent to drought, and repels birds, animals and insects from the crops. It is a very short duration crop, and starts flowering during the first week of November. The Regional Research Centre of OUAT, Semiliguda, has improved selection of *alsi*, from the existing cultivars.

Plate 92: Ragi, multi finger indigenous variety, Lamptaput, Gadaba

Plate 93: Fox tail millet, Kashipur

Plate 94: Suana, Laxmipur

The castor plant is the tall, native type. The tribes utilize castor oil. The castor oil plantations are found more in the Rayagada area. The plant also serves the border fencing on slash and burn cultivated fields.

Gangwar and Ramakrishnan (1989) reported some lesser known plants of food value cultivated by the Khasi tribes. *Digitaria cruciata var. esculenta* is the cereal crop and a source of compost in 90 per cent of the villages. It is often intercropped with potatoes. They found *Perilla ocimoides*, a seed crop cultivated in all villages and *Flemingia vestita*, a legume crop in 48 per cent of the villages. Arunachal Pradesh tribes grow *Discorea* specie under tree stumps to gain support; along with it a mixture of *Amaranthus viridis* and *Chenopodium ambrogioides*. The former, Amaranthus crop yields leaves and the Chenopodium yields leaves and seeds and are utilized for food. *P. ocimoides*, *Panicum miliaceum*, *Setaria italica* and *Coix lacryma crops* are sown in March and *Fagopyrum tat aricum* is sown separately in July. *F. vestita* contributes up to 250 kg N/ha. in a year.

There are also such prospects on Koraput hills to isolate some more legume food crops.

The life support species of plants, as well as the famine saviour plant species, are the most vital factors in the lives of tribals. These groups of plants, particularly the life support species, are also most underutilized plants. Research efforts to improve productivity of these groups of plants are being neglected. This is so because poor people have always been neglected, do not form a part of the policy decision making process. Besides, any useful research work done by individuals or institutions in this sphere, usually remains unrecognized. Both these factors are responsible for the current state of affairs. The productivity of these plants has to be improved. The sequence of gradual maturing food producing plant species should be encouraged so that the tribal families have continues food supply system.

The ragi millet has at least 3 genotypes based on the length of the crop at harvest. The earliest ragi varieties are those which are harvested by Dussehra (September-October). The late ragi varieties are harvested by December. In between are the medium varieties of ragi. These varieties synchronize well with the food demand of the family, as well as the land and moisture availability. In this way, the tribal farmers are able to utilize their experience and knowledge to optimize the immediate family food requirements as well as reserves for the future. Thus, the land size and type are allotted accordingly.

Prof. M.S. Swaminathan (1996) had said in the Madras Declaration, 1996, that there is a need to move away from a 'favourable area' and 'favourable crop' mindset, and to enlarge the contents of the food basket by including the excluded in terms of underutilized plants and neglected ecosystems. If implemented, this will be of immense value to the tribals in times of acute food shortage. The existing food gathering and food production system needs careful attention, particularly during the phase when tribal agriculture is in transition; because some alterations in the system can result into serious consequences for life and living.

Scarcity Period and Food Resources

In terms of intensity and lengths, tribes have differing periods of food scarcity. This is particularly evident among the hill tribes, who have to live with depleted natural forests. This is bound to happen during transition from a forest dependent economy to agriculture.

During the lean periods, when resources from agriculture and the animal husbandry system are scarce, particularly for the poorer sections, wild fruits help them tide over the crisis.

Up till the time of harvesting the long duration paddy, and even afterwards, families use their food resources for consumption at home, as also for marriages, guests and festivals. By mid and late summer, however, they become careful about their food resources. Families keep seeds for future sowing carefully and retain more quantity than required as per seed rate so that in the event of failure of germination the family can still meet the seed requirements.

To meet their expenses in times of scarcity, young men travel in groups in search of work. Although they travel to places far from their village, they usually do not go beyond the limits of the district. Women from the village also work in nearby road or plantation works.

In very acute scarcity food conditions, when all resources to purchase food from market are exhausted, they start utilizing non-conventional food resources. This, in itself, is an indication of the seriousness of the problem.

In 1825, Stirling wrote about the eating habits of the Saora. He wrote that,

> A great part of their subsistence is derived from the roots and produce of the jungles. The flowers of *Bassia latifolia* and the Keora (*Pandanus odonatissimus*) yield them an intoxicating liquor, in lieu of rice they consume the seeds of bamboo, a very heating and indigestible food; the wild yams, arums and other roots furnishing a nutritious, and not unwholesome substitute for bread; for a desert

they have the wild mango, the fruit of the Bela every where is abundant, and the seeds of the *Bauhinia racemosa*, served up on the large ribbed leaf of the Ravya (apparently a species of Dillenia), which answers the purpose of dish.

This has mostly remained unchanged since then.

Tanku Pej (Mango Kernel Pulp) and Tamarind Seed Pej

The Tanku pej and or gruel from fried tamarind, which have been soaked overnight after their outer coat has been removed, is what the poorest of the poor consume during times of acute shortage. These usually fall during the Landi and Ashad months (in June and July, up to 10 August) in the Kondh villages. The Sabara Bondo, Didayi, Kondh, Dongria Kondh and Koya also use mango kernel pulp as food.

After it is thoroughly cleaned and repeatedly washed in the running streams, mango kernel is dried in the sun. Women process this at home during the late summer months, and a gruel is made from it during July and August. In the process of preservation, the mango pulp perhaps gets contaminated and is suspected for some deaths among the poor tribes in the hills. They are often diagnosed with cerebral malaria and Japanese encephalitis, etc. The tribals are advised against rearing pigs because it is believed that they act as the hosts for the encephalitis.

Mango pulp is normally consumed from 15 June to the first week of September.

Sago Palm Pith Pej

In certain places, the dead Sago palm stump is broken and the pulp is eaten as gruel, with bread by the tribals. The two food substitutes mentioned here are usually consume during time of severe food shortage. The Kondhs and Sano Poraja tribes often use these food resources. The tribes who have *mahua* trees, use dried ripe *mahua* flowers as food during these scarcity periods.

Bamboo Shoots

In August and September, if bamboo is easily available, bamboo shoots serve as the staple food for the family. Not all bamboo varieties are edible but the tribals know the difference and are aware of the specific varieties that are edible. A person can collect up to 150 bamboo shoots in a day. Since these could contain the fatal hydrocyanic acid, the

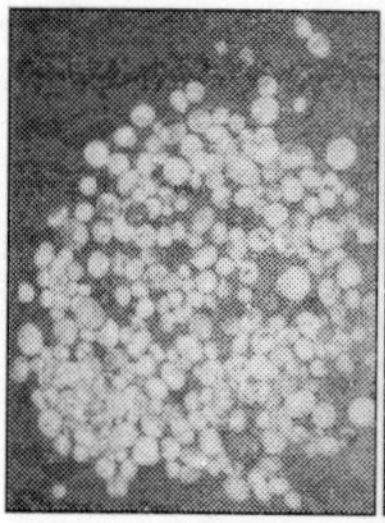
Plate 95: Boda Chatu

Plate 96: Sano Bali Chatu

Plate 97: Mania Chatu

Plate 98: Bodo Bali Chatu

Plate 99: Kendu fruits

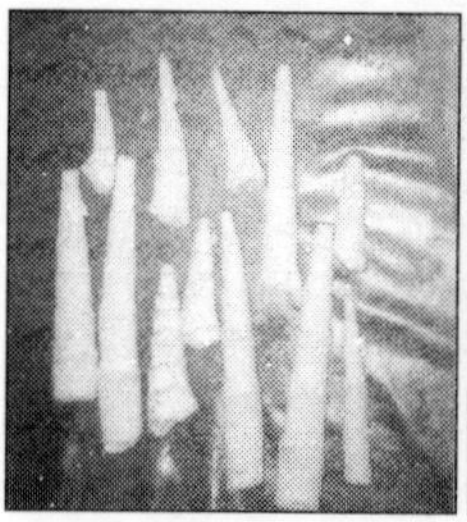
Plate 100: Bamboo shoots

Plate 101: Kanta Kuli

Plate 102: Char Kuli

Plate 103: Mango kernel for gruel

Plate 104: Wild mango pulp and juice processing

Plate 105: Jam koli

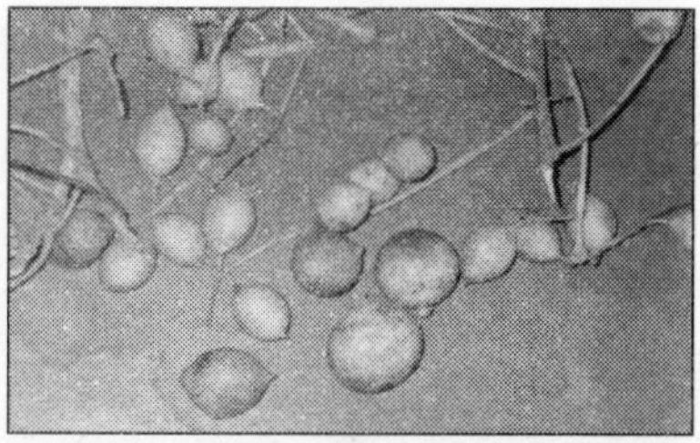
Plate 106: Kusum koli

Plate 107: Dumburi

Plate 108: Podaei koli

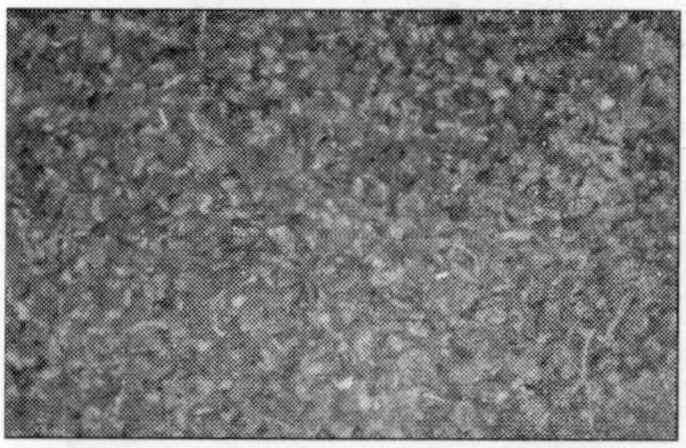
Plate 109: Mahua dried flowers

Plate 110: Siadi fruit (seed)

Plate 111: Jackfruit tree

Plate 112: Collecting wild mango

Plate 113: Sago palm tree

shoots are cut into thin, long chips, with longitudinal cuts, and immersed in water for 48 hours. The water is decanted and refilled at 24-hour intervals. Sometimes, they boil the bamboo shoots in water and preserve them by drying them in the sun, as additional food reserves for the family.

Barik (1997) has studied 10 varieties of shoots from 10 Bangladesh bamboo. *Melocanna baccifera* is the most easily available and the most common and edible bamboo type for the tribals. They have advised in favour of annual selective harvesting of bamboo shoots.

Tewari (1988) has reported bamboo as poverty alleviator for tribal areas of Madhya Pradesh. The bamboo shoot has been described earlier as the most vital food available to families during acute scarcity in the annual food cycle.

Certain green leaves like 'Gudgudia sag', 'Katabhaji sag' are also used as food during these months. The former grows in water-logged areas. The Ghir kanda or tuber from forests is another food item available from November to April to the Kondhs.

In the plains, the pumpkin ripen in mid-August and sometimes even the cooking-type plantains and green maize are available from the backyard. Bondo tribe cook pumpkin with cereals for family food during scarcity period. These are sold in the weekly markets and cereals are bought for use at home. On the hills, fox-tail millet and early suana are ready for harvest in the second week of August.

The small patches of suana on the hills are harvested with hand knives. This is dried on flat stones, although care has to be taken because of intermittent rain. Used as gruel, this suana lasts for a week or two. Then, in September, the medium duration suana appears followed by kangu, early varieties of ragi and the early variety of upland paddy, and long duration of suana, along with maize. After November, the beans and green *arhar* pods start appearing.

After a village ceremony they start eating roasted beans and *arhar*. So a process of gradual harvesting and marketing the produce appears to be in place. But only those who have some land and have remained healthy and fit to work in the fields during pre-monsoon period can make use of it.

The Bhumia tribe of Kundra and Boipariguda blocks eat *kulathi* (horse gram), a leguminous grain, during times of food scarcity. However, they deny consumption of the piths of the sago palm bark.

TABLE 7: ANNUAL FOOD CALENDAR FOR KONDH TRIBES OF LAXMIPUR BLOCK

Kondh months	English months	Food (main)	Season	Food	Work
Base	May	Mandia pej (Tentili pej)	Mango	Koida and Borda sag	–
Landi	June	Mango kernel pej (Tentili pej)	Mango	Koida and Borda sag (not after Gundicha)	Mushroom, Badi sag and vegetables
Ashad	July*	Mango kernel pej	–	–	Mushroom, Badi sag and vegetables
Bandapan	August*	Mango kernel pej	–	–	As above
Barsi	September	Suana pej	–	* (Rodika kanda, Sirkanda kanda from forest)	Mushroom, Bodi sag and vegetables
Dashara	October	Suana pej			
Diali	November	Mandia, rice, Suwan, kangu, maize pej	–	Rani *kanda*, simili kanda, cultivated tubers, Ghir kanda from forest	–
Panda	December	Mandia, rice Suwan, maize pej	–	-do-	–

(Contd.)

TABLE 7 (*Contd.*)

Kondh months	English months	Food (main)	Season	Food	Work
Paus	January	Rice, mandia pej	Kandula, other lentils	Ranikanda, simili kanda tubers	–
Magh	February	-do-	-do-	Rani kanda, simili kanda tubers, *mahua* flowers early	–
Phagun	March	-do-	Jhota (creeper flat beans)	Ghir kanda, *mahua* flowers	Tamarind collection
Chait	April	-do-	-do-	Ghir kanda, Borda sag, *mahua* flower ends	-do-

Note: *food scarcity months; sag means leaves used as vegetables; kanda means tubers.

The following crops and tubers are collected from the forests for food.

Non-cultivated leafy vegetables of the sag group are mostly available during the rainy seasons:

Gudgudia sag	Katabhaji sag	Bhaji Boda sag
Sati sag	Koida sag	Borda sag, etc.

For the tribal groups of Jeypore, the following tubers are available as food from the forest during the summer and rainy seasons:

Saronda kanda	Kasa kanda
Chiting kanda	Pit kanda

One needs to be careful of Kolia konda, which is intoxicating.

The following types of the berries are usually available during the summer months:

Char koli	Kanta koli	Podaie koli
Moir landi	Siadi koli	Kendu fruit
Amla koli	Kusum koli	

The Koyas of Malkangiri consume dried *mahua* flowers with other available grains as gruel during the scarcity period. They dry the ripe Kendu fruits during summer and preserve them. These are immersed in water and reconstituted before being eaten. This also helps them during the time of scarcity.

The Wild Roots and Tubers Consumed by Koya Tribe of Malkangiri

The roots and tubers that they eat usually belong to creepers and run deep into the soil. Some are found in thick forest, under shade, some others are grown at home, while some are found in ponds.

The forest remains accessible to the tribals only during early rains, when they can go in search of rhizomes, bulbs and roots. As the rains advance, however, the forest floor gets thick with vegetation, hampering the search for roots.

Kirsamati: This is a creeper. It likes to be under shade. The tubers are about 6 ft long and grow in semi-circle under the ground. It is available from late August to mid-October. The tip of the tuber reappears on to the surface during Dashara; so Koya women say Kirsamati has come up to see Dashara.

Phadmati: The leaves of this creeper are round. The tuber is 2 ft long and is available in the forest as above.

Dorimati: The creeper can be 50 to 100 ft long. Its tubers are 5 to 6 ft long and 3 to 6 inches wide. The leaves are thin and long. It is available in June only.

Pendumati: This is a thorny creeper, whose roots spread out from the bulb. The leaves are oblong and it is available from late May to June.

Keskamati: This creeper has round leaves and a white bulb. It can be preserved at home for one year. It is available from late August to mid-October.

Kaimulmati: A creeper, it has round leaves and a black bulb. It is available as above.

Krosodamati: This is grown at home and is also called Similikonda. It is available during June.

Mamamati: This creeper appears after the first rains. Its leaves are like Nangolkonda, while its tuber is 3 to 4 ft long and 6 in. wide. Sweet in taste, it is quick to boil. It is found during late April to May.

Durupamati: It is found in water all round the year. There are 3 types. The leaves are round. One has a small bulb with tiny bulbs attached to it, the other has a larger bulb, and the third type has small size bulbs. The fried bulbs are tasty.

Tamarmati: It is found in water round the year. This is the Purnikonda of ponds. It is edible.

Kukadakanda: Tubular in shape, this is found in water and is very sweet.

The last three are used during June-July, when there is scarcity of food at home.

During acute food scarcity period, tubers like Kirasmati, Keskamati and Kaimulmati are available in the forest, along with less palatable Ahadmati, Krisagonda, and many more used for food.

The Baunso chaulo are the seeds of the Kodikia variety of forest bamboo. The Koyas and Sabaras cook it during times of acute shortage of food. The fresh Phadmati and Keskamati are intoxicating. They need to be immersed in water for at least 24 hours, before being cooked (Kornel, 2002).

During livestock and poultry epidemics, when a sizeable number of animals die at home, the tribes who eat dead animals are Sano Poraja, Dongria Kondhs, Kondhs, Bondo and others. They cut and dry the carcasses, before using them. Eating dead animals is another indication of the extent of the problem during times when food is at a premium. The anthrax epidemic has struck repeatedly in the Narayanapatna and Dasmanthpur blocks, affecting both the livestock and human beings, specially among Kondhs. It is as much a measure of their ignorance as it is of their desperation that they continue to eat dead animals, which are the very source of the disease.

The Mahua Flower and Fruit as Food of Koyas

Before the new season starts, the Koyas celebrate 'Bhimulu pendule' by collecting and eating the *mahua* flower. The *mahua* trees have ownership, except for few in the forest. Before the collection of *mahua* flower, the floor around the tree is thoroughly cleaned. Only the flowers that fall to the ground are collected.

The Koyas worship the *mahua* tree. Almost all the festivals, ceremonies and treatment of diseases by Sirha, are performed under the tree. It is also called the 'mullo tree', which means origin or centre. The Koyas consider the *mahua* tree the centre of their culture and life. The Koyas consider the *mahua* wood as holy for the pyre.

When the *mahua* tree is flowering, the women and children go out to collect the flowers at 5 a.m. They usually return either at 11 a.m., or at 5 p.m. The fresh flower is either dried for sale or is preserved to be used as food.

PRESERVATION METHOD OF *MAHUA* FLOWER

The flowers are dried in the sun for two days, and more flowers are added to the collection everyday. Stored in a heap inside the house, the flowers are beaten on a clean bamboo mat and the crown of the corolla is removed. After being cleaned thus, the mahua flower is ready for preservation.

An open-ended, cylindrical shaped bamboo basket is knitted. In it, a layer of siadi leaf is closely packed against the inner wall. The two are then stitched together to form a closed bag. The dried *mahua* flowers are then packed layer by layer, inside the bag, and pressed well. A thick layer of siadi leaves is placed on the open end of the basket. Siadi string is tied to the rim of the basket and pulled, to try and seal the basket as best as possible. Thus sealed, the basket is now moisture and insect proof. The *mahua* flowers inside can now be preserved up to two years and are ready to eat in six months.

The Koyas subsist on the *mahua* flower during times of acute food shortage. But the flowers and cereals are never cooked together. The Kondhs of Kolnara, however, cook the *mahua* flower with bailo seeds. Elwin (1950) wrote about the Sabaras, that, 'They make a sort of pudding out of the corollas of *mahua* flowers and tamarind, which they boil together in a little water all night. They also roast corollas with *amaru* seeds in a dry pot and make it into balls, which the children love to eat.'

The Koyas boil *mahua* flowers in water for 2 to 3 hours, and sometimes even up to 6 hours. It is ready to eat on the following day. The other *mahua* flower dishes are:

(i) The bendagunda seeds (kaunria plant) are heat treated, and ground to flour. Then it is mixed with the fried *mahua* flower and eaten.
(ii) The treated bamboo shoots are mixed with dried *mahua* flower. Koyas eat this with relish.
(iii) The *mahua* flower is fried in 'tolo oil' and is served to the family.
(iv) The ripe gomhar koli is mixed with the *mahua* flower and cooked. This also makes for a tasty dish.

The Mahua Fruit (Tolo) as Food

The *mahua* fruit is eaten raw and its seed and skin are discarded. The ripe fruit, tolo, is used to make wine. The flesh of the fruit is boiled and the decanted water is thrown out. The fruit is then fried or mixed with rice and cooked. The tolo fruit is also mixed with the *mahua* flower to make another dish.

Siadi (Bauhinia spps)

This creeper is found in the forest. Some siadi creepers, which are hundred years old, or even more can be seen in the religious groves of the tribes. The siadi pods, with dry seeds, are available from March.

The seeds are roasted and an adult can consume between 50-100 seeds in a day. The Koyas believe that eating too many seeds can cause abdominal pain but the Dongria Kondh think it is quite safe.

Mushroom from Forest

The tribals, especially the women, are very good at distinguishing the edible species of the naturally growing mushrooms that are locally available. Dug out from the ground, the 'Bodachatu' (*Gastromycetes*) is round and has a thick cuticle. Usually the cuticle is black but when it is dug out from a sandy soil, it is white in colour. Found in June and July, its pulp is soft and black. There is a very small, tiny brownish white mushroom, which grows naturally in shaded places and is called 'Khudachatu' (*Termitomyces group*). This is of two types, with one being slightly larger than the other. It is found in July and August. The most common mushroom is called 'Srabanachatu' or 'Maniachatu'. This large mushroom is found in abundance in the sal forests during August and September. Tribal families collect and sell most of these edible mushrooms in the weekly and daily markets. They receive good price, in the early part of the growing season. There is a long list of naturally growing mushrooms known to Koyas as edible, as also on Bondo hills and to Dongria Kondhs. This needs documentation. The non-edible mushrooms are called 'Bayachatu'.

Harsh et al. (1996) studied edible and medicinal fungi collected by families in Madhya Pradesh forests. The main edible species were *Astraeus hygrometricus, Calvatia cythiformis. Geastrum fimbsiatum, Geastrum triplex, Lycoperdon pusilum, Mycenastrum corium, Podabrella microcarpa, Termitomyces helimii and Scleroderma rudicans.* This holds good for Koraput district.

The Famine Saviour Plant System

Grigson (1938) wrote that, 'Famine has never been a problem in Bastar, as the Marias and other tribals have always been able to draw half of their food supplies from the innumerable edible products of the vast forests'.

In terms of intensity and length, tribes have differing periods of food scarcity. This is particularly evident among the hill tribes, who have to live with depleted natural forests. This is bound to happen during the transition from forest dependent economy to agriculture.

The very common fruit trees that are found in and around the tribal villages of Koraput are tamarind (*Tamarindus indica*), mango (*Mangiferra indica*), jackfruit (*Artocarpus heterophyllos*), sago palm (*Caryota urens*), jamun (*Syzygium cumyni*), kendu (*Diospyrus melanoxylon*), *mahua* (*Madhuca indica* or *Madhuca latifolia*), toddy palm (*Borassus flabellife*) and a few trees, like wild date palm (*Phoenix sylvestris*). The tribals still respect trees, which provide them food and are witness to all their celebrations, religious functions and other happy occasions at home and in the village. As a result, they never cut down trees. This applies even to the new saplings. It can be said that they are very conscious of fruit as well trees. Invariably, most households among tribes found in the plains, who have a good backyard around their houses, have various fruit trees and bamboo plants around its boundary.

The Dongria Kondhs of Bisamcuttack and hill Sabaras have a special attachment to fruit trees in their life. They grow various fruits, like jackfruit, tamarind, mango, orange, and banana, etc. The Kondhs of Phulbani, Gadabas of Lamtaput, Bhumia of Boipariguda, Bondos on the hills of Khairput, Didayis of Kudumulgumma have a special liking for jackfruit trees. The Bello (*Crateva marmelos*) was one of the important fruits of Sabaras during the eighteenth century. Among the Koyas of Malkangiri, Gonds of Raighar, Kondhs of Kolnora block and Sabara of Gunupur, the *mahua* tree has special significance.

During periods of severe shortage of food, it is the mango tree in

Koraput district and the jackfruit tree in Phulbani, which help the poor tribals to tide over the crisis. Fruiting of the mango tree is largely dependant on the bi-annual flowering phenomenon. On occasions when the mango tree does not flower, the tamarind tree helps the situation. In March-April, the *mahua* fruit helps to improve the situation. *Mahua* is an area and culture specific plant.

Tamarind trees: An evergreen tree, it grows to a every large size. Its compound leaves consist of many leaflets and grow so densely that they cast unbroken and pleasant shade on the ground. Its leaves are light green when young and dark green when old. The tree flowers twice a year and its pods are harvested during summer. Irregularly swollen, the pod is curved and brown in colour, containing several seeds that are covered in fibrous pulp. Besides being consumed, the fruit is an important article of commerce and is exported. At times when food is scarce, the seed is eaten after proper treatment. Young leaves of the tree are also used as vegetable.

When ripe, the tamarind is picked for sale by villagers, and finds a good readymade market in neighbouring Andhra Pradesh and elsewhere in the south. The separated seeds are also sold in western part of the country. The seeds are fried and cooked as gruel, which has also been mentioned elsewhere. Sold between Rs. 2 to 2.50 per kg (2001), it is also bartered for salt. In a satisfactory crop year, as much as 100 tons of tamarind pulp is exported from Koraput district alone. When the crop is not good, however, production goes down to about 50 tons.

Mango trees: Mango is the next important fruit tree. Mangoes are eaten during summer after an elaborate ceremony, which has been discussed elsewhere. Mangoes of the region are available in a great variety, differing in their colour, size, smell and taste. During summer, families, with children, can be seen spending most of the day under mango trees, collecting and eating mangoes. During most celebrations, marriages and construction of a new house, etc., the tribals use branches and leaves from the mango tree.

Fresh mango juice, with pulp, is the main source of calories during the summer. Families in the hills preserve excess mangoes in the form of either dried juice or dried pickle. Although the fruit of the naturally growing local varieties is small, there is a great diversity in terms of colour, smell, pulp and taste.

Plate 114: Village Community mango Orchard, Phupugaon

Jackfruit tree: The Nandapur, Lamtaput, Bondo hills and Niyamgiri belt of Bisamcuttack, as also the Tangniguda centre of Boipariguda areas, are the rich sources of jackfruits, which are harvested during early part of the monsoon. Unripe jackfruits are also consumed as vegetable. The large, edible fruit can weigh anywhere between 3 to 20 kg and is sweet in taste. It is as attractive to human beings as it is to the wild bears. Its seeds are also edible and are cooked before eating. In terms of size, colour, shape, type, smell, sweetness and taste, the fruit is available in great variety.

Jackfruit of Koraput finds good markets in Nagpur, Agra, Kanpur, and Bhopal, where the unripe fruit is used as vegetable and pickle and goes for a good price. Tribal families also use the unripe ones as vegetable, while the ripe jackfruits are consumed as fruit. The routes to export are Bisamcuttack to Nagpur by railways and from Dasmantpur area, south of Boipariguda, by road to Nagpur. The Bisamcuttack ones are all the early fruiting type, whereas the Phulbani district are mostly late fruiting variety. As a result, the former has better marketability and price. It is estimated that 40 to 50 full truckloads of unripe jackfruit exported annually from Tanginiguda centre alone to north India (2000), apart from local sale and consumption.

Jamun tree or Jambokoli tree: This is also a very large tree, whose fruit is juicy and has a single seed. The skin and pulp of the fruit are green at first. Gradually they turn purplish and end up being dark purple. There are two types of jamun, one which has very small fruit and the other, whose fruit is large. The large fruit is up to 15 mm in length and 10 mm in breadth. The fruit is very tasty and colours the mouth a deep purple. The branches of the plant find a place in most of the religious functions of the tribal families.

Sago palm (Caryota urens): This grows up to 10 m high and can even go up to 30 m. The leaves are about the largest of any Indian palm, while the leaflets look like a fish tail or fin. As a result, it also gets the name Fish tail palm. It is indigenous to India. The plant is very profitable. The fibres from the leaves and flower stalks are used to make durable ropes; from the pith of the palm, a type of flour is obtained and is a source of food during the scarcity period for the Koyas, Sano Poraja and other tribes as said earlier. Its sap, fermented or otherwise, makes for a refreshing drink. Ceremonially, people plant a lot of small saplings all around the house, especially in the bathing area and where utensils are washed. The sale of the juice from a fully grown sago palm tree brings in a good amount of cash for the tribals.

Mahua tree (Madhuca indica or M. latifolia): The *mahua* tree is central to the lives of all tribes of central India including Umerkote Gonds and Malkangiri Koyas. The tree is held to be so sacred that the Koyas do not climb the tree, cut it, or even break twigs from it. The tree is also called 'Mullo', which means origin or centre. As a result, all important ceremonies of life among the Koyas, including the treatment of the sick, take place under the mahua tree. The mahua flower serve as food, especially at

Plate 115: Flowering mahua tree

food is scarce. Surplus flowers are sold, as they are in times of need too. They eat the *mahua* fruit, while the kernel, called tolo, is used to make oil. This oil is their only cooking medium. Various dishes are made from *mahua* flowers and fruits. The flowers are fermented to produce liquor. As a result, the local alcohol manufacturers use up a large number of *mahua* flower.

The flowering starts from February end and continues up to April end. The fruits are picked from second week of May to the first week of June. A detailed study about *mahua* trees has been published by Ram Prasad (1993).

Kendu tree (Diospyros melanoxylon): This is another economical tree for them. The sale of dried Kendu leaves, which are used by the bidi industry, is an important source of earning for the tribals during summer. The large trees bear plenty of round fruits, while its pulp is very sweet and a rich source of energy and vitamins. The Koyas dry the fruit for future use.

Similarly during summer, they collect sal seeds (*Shorea robusta*) from the forest and sell it to the traders. The seeds are used in the oil extraction industry.

Toddy palm (*Borassus flabellifer*) is also known as Tal, Tar and Tad. It is grown in Malkangiri area among Koyas. This palm may reach 20 to 30 m in height. Its fan-like leaves are 1 to 1.5 m in diameter and have between 60 to 80 segments. Between 60 to 100 cm in length, it is rigid, folded along the midrib and has a stout leaf stalk (Santapu, 1966). In the toddy rich areas of Podia to Konta, whole villages of Koyas camp out in the toddy groves from March till June in a state of semi-intoxication. They eat solid food in the evening but sometimes go without it for three to four days. The sap, which appears between November and March on the male trees and between April to June on the female trees, is tapped by cutting the ends of the flowering spadices. The vessels to catch sap are generally thick bamboo tubes or earthen pots. The juice soon ferments to make an intoxicant. Fresh juice is mixed with some leftover from the previous days. Female trees bear fruit in April-May, and their pulp is eaten. They also eat unripe seeds, seedlings and also roots.

The Koyas sacrifice a fowl or a goat at the foot of the first tree they tap.

Wild date palm (*Phoenix sylvestris*) is found among Koyas of

Malkangiri, and Sabaras of Gunupur-Paralakhemundi. The tree is grown on paddy field bunds, house boundaries and on hill slopes by Sabaras. The tree yields a sugary sap that can be fermented into liquor, or concentrated into sugar. The wild date possess a large stone in the centre, surrounded by a tough skin and little flesh. The ripe dates are collected to eat by Koyas and Sabaras.

Karanj tree (Pongamia glabra): This tree grows in Gutob villages. It is a large and hardy tree, whose dried fruits are used for collection of oil by primitive press and hot sun effect. It is used as hair dressing oil and body massage oil in winter. The oil is yellow and viscous. Known for its medicinal value, it is used to treat skin diseases.

Muninga tree or Drumstick tree (Moringa pterygosperma): Its tender leaves, as well as the long and narrow fruits are cooked as vegetables. It is of great value to few tribal groups and the prospect of growing new improved varieties is high.

Bamboo plants: *Dendrocalamus strectus, Bambusa auriculata and Ouytenanthera nigrociliata* spp. are the types found in the area. These plants find a multifold use by the tribe, and are grown carefully around the boundary of the house. They are used for thatching roofs and for making common household materials, like mats and bins, etc. After they have been cleaned of hydrocyanic acid, young bamboo shoots are used as vegetables. They have a large collection of bamboo species in their villages, like: Sundarmania (beautiful), Buta-all (bushy), Kanta (thorny), Khupu, Bariaha (wild boar), Patikial (beam), Tapu and Dongar, etc.

Commonly found almost everywhere in the area, Sundarmania is a long and fast growing type and is used to make household and utility things. Its young shoots are not used as food. They are really very beautiful, with long green leaves, clean and shining bamboos and grow very tall rapidly.

The other minor fruit trees of importance are Ram fala (*Annona reticulata*), Sita fala (*Annona squamosa*), Dumbri (*Ficus carica*), Banana (*Musa paradisiaca*), and others. But of late, Bhalia fala or Cashew (*Anacardium occidentale*) is being planted in large areas of the Nandapur, Padua, Kundra, Malkangiri and Lamtaput villages, both by the government and privately. The nuts are sold to factories located at Jeypore and the apples are being eaten as fruit. They are not aware of the process to make alcohol from apples.

A people's participatory appraisal method of survey was conducted by Breinholt and Kornel, covering 100 villages of ILDP during 1993-4. This showed that there were 18,968 fruit trees in 100 tribal villages of Kundra, Jeypore, Koraput and Laxmipur blocks of Koraput district.

The trees of importance that were recorded were local and grafted mangoes, jackfruit, tamarind, cashew, citrus, guava, coconut and custard apple. There were more trees of the local mango varieties, 7,988, than any other. This was followed by 3,290 jackfruit trees, 1,888 grafted mango trees and 1,446 guava trees. Cashew had been planted over 1,213 ha.

On analysing the data of fruit trees, it was observed that the highest number of local mango and jackfruit trees are found in Laxmipur block, followed by Koraput. Tamarind trees were equally distributed in almost all villages of Jeypore, Koraput and Kundra blocks. The coconut trees were found in Jeypore block, while in other areas there were hardly any. Grafted mango was known only in 72 per cent of the villages in the Jeypore block and very few in Laxmipur. Similarly, custard apple is found in very small numbers in Jeypore and Koraput. However, it is very common in the Malkangiri, Bandugaon and Narayanapatna areas.

Fruit trees such as tamarind, jackfruit and local mango will remain important for all the tribal villages located on hills. Large-scale plantations of mango and coconut in 1994, however, did not find the tribals' approval. Perhaps this can be attributed to lack of awareness about the trees and the skills involved in planting and looking after them. In terms of priority, tamarind is at the top of the list of fruit trees, followed by jackfruit, guava, citrus and local mangoes. Cashew, a recent addition to the area, is catching up quickly. The food security of tribal families' has improved further due to sale of harvested cashew nuts in the market.

Cultivated in the otherwise unused gravel uplands, cashew is becoming extremely popular because of the profit it offers.

The cashew plantation is strongly supported by the upcoming cashew industry, which has about 28 processing factories in Jeypore alone.

The reduction of poverty is most time consuming; instead quick results shall be sought. At the stage, when the poverty programmes is to be evaluated for impact, the failures shall be leveled against liquor and thereafter *mahua* trees will be eliminated from the system. The *mahua* tree provides food to people and is well acknowledged in tribal life. It needs to be respected.

Slash and Burn (*Podu*) Cultivation System

Locally known as 'Podu', slash and burn cultivation system is also known as 'Jhum' and shift cultivation. This agricultural practice is particularly followed in the hilly areas of Koraput district, where only limited land is available for agriculture. Ruthenberg (1980) defined 'Shifting Agriculture' as a term for a system that alternates between cultivating a plot of land and then leaving it unused for a certain amount of time, to allow the soil to recuperate.

Shifting cultivation is often a technology of expediency, because it will work better under a given set of conditions than in any other system available to the practitioner (Spencer, 1966).

On the other hand, it is a land-use system that may belong to the past, but is still carried on because the transition to a more productive land-use system demands more favourable price relations, knowledge, capital and time. Bhupinder Singh (1987) wrote,

> There is a lively debate over shifting cultivation. Views have been expressed to the effect that this mode of cultivation is injurious to the eco-system and, on the other side, that is a tribal way of life not to be interfered with lightly. Though it is not possible for me to take the position of supporting Jhum on the ground of its contribution to the eco-system (as some students claim), I cannot help observing that a change in the lifestyle of a community is not a matter of technology alone and sociological and human dimensions are weighty. Further, shifting cultivation is characterised by so many ethnic and regional variations with in-built safeguards that its condemnation out of hand would be imprudent. The students of podu would do well to unravel all its features including those which make it conservation-oriented. Above all, energies should be concentrated on evolution of viable alternatives, since few are so far in sight. Horticulture, rubber, coffee, tea, etc., plantations have been suggested and in some places tried as alternatives. In the State of Orissa, in the middle of seventies we had laid out thousands of acres of horticultural plantations, but most of these languished for want of proper marketing arrangements. Tripura has achieved some success in rubber plantations and Andhra Pradesh in coffee plantations. But there are snags. High cost schemes have to be

abjured. There cannot be escape from the shadow which lies between the concept and reality. Hence, not only are the fine and dedicated men and women required for originating concepts, but equally determined and perceptive minds are required to master the art of implementation.

Ramakrishnan and Gangwar (1989) studied the village eco-system function of the Khasis in Meghalaya and opined that the land has been decertified around Cherrapunji, as the traditional slash and burn agriculture has been replaced by plantation systems. Apart from mixed plantation systems, with areca nut, orange, jackfruit, bay leaf, black pepper and betel leaf, broom grass, thatched grass and bamboo are also raised in the village. With labour as the major input, generally, all the systems are economically viable and have a higher energy efficiency.

Sarma et al. (1995) studied the effect of shifting cultivation in Assam. The study was conducted on land with 20 per cent slope, with trials involving traditional and improved cultivation practices. The highest yield was recorded when pineapple, turmeric, ginger and arum were grown in the border and rice, maize, sesame and cotton were sown across the slope.

Mohapatra (1992) has described Podu, as an ecological hazard. Behura (1990) has described the shifting cultivation as a cause for environmental degradation.

The saddest part of podu cultivation is the felling of the valuable timber trees, only to find that plot of land, with more boulders and stones than soil, is not ideal for cultivation. This is common in the Gajapati district, where the crops are limited to jowar or castor. Obviously, the yield from such plots is meagre, when compared to the loss of trees. Perhaps greater awareness may have saved the trees and wider use of fertilizers may have contributed to a higher yield. I am inclined to believe that lack of knowledge of any alternate system is responsible for the destruction of these trees.

Mohanty (1986) has described the shifting cultivation employed by the Juang tribe of Orissa. The Bondo, Didayi, Koya, Kondh, Gadaba and Saora tribes of Koraput district have been classified under tribes practising podu or shifting cultivation in Orissa. The shift cultivation fields are known as:

1. Juang of Keonjhar hills - Ekan, Guda, Nala

2. Bhuiyan of Bonai in Sundargarh	-	Kamar, Biringa and Guda
3. Saora of Ganjam and Koraput district	-	Bagada or Sarban
4. Dongria Kondhs of Rayagada	-	Harh
5. Desia Kondhs of Ramakhal	-	Rama
6. Bondo of Koraput district	-	Livang or Kunda chasa
7. Koya of Koraput district	-	Lanka padseenad

The Bondo and Didayi communities mostly live on shifting cultivation. About 70 per cent of Kondhs and Saoras still live on shifting cultivation and about 19 per cent of Gadabas.

Pattnaik (1988) gave a good account of the shift cultivation scenario in Koraput. According to him, in Orissa, shift cultivation is being practised over 23,72,705 ha of land. About 91 per cent of this lies in southern Orissa alone and 9 per cent in northern Orissa. A rough estimate shows that 6,47,000 persons in south Orissa and 59,412 persons in north Orissa practise this form of cultivation. Roughly 7 per cent of tribal population carry out shift cultivation. In recent times, these figure have gone up.

Under *Podu*, a plot of land is cultivated for three years and then left alone for a certain amount of time, to rest and recuperate. A few years ago, this rest period used to be between 10 to 15 years. Owing to the increasing requirements of a growing population, today, this rest period is no more than three years. In some cases, it is even less.

Owing to the higher labour input and the crop mixture used, the economic efficiency of slash and burn agriculture followed by Garos was higher than that of the Khasis or the Mikirs (Maikhuri and Ramakrishnan, 1993). The Garos have more efficient animal husbandry systems compared to the Khasis.

A comparative analysis of the agricultural and animal husbandry systems carried out by two tribes in Arunachal Pradesh, the Sulungs and the Nighis, shows that the traditional agriculture and animal husbandry of the tribal people are found to respond to both human and environmental needs. If labour costs are considered, the financial investment in podu is high, but since the labour output is geared to the maximum utilization of family labour, it does not have actual financial costs (Gangwar and Ramakrishnan, 1987).

Plate 116: Firing the forest, Rayagada

Plate 117: Stony land after firing the forest, Rayagada

The Podu cultivation on hills is administered on community basis, with full discipline. There are hardly any disputes or claims over particular plots of land. Perhaps this is so because, owing to physical limitations, families are unable to clear the bushes or prepare the land in the short time available before the outset of the monsoon rains.

For decades there have been several recommendations to stop Podu, but little progress has been made in enforcing this.

Rotation Systems

The shifting cultivation, as a form of 'rotational system' is generally the 'classical long-fallow systems' pattern, classified by Miracle (1967). Here crops are grown for 2-4 years and then one decade of fallowing. Nowadays, the prevailing trend is both for medium term fallow period (3-5 years) and short term fallow period (1-2 years) in and around Laxmipur area.

A good account of shift cultivation in north-east states of India has been given by Pandey, (1996). He has reported fallowing period between 2 to 10 years.

Cropping Systems

It is always exclusively carried on as farming with annual and biennial crops. Cereal crops and, at times, legumes are grown; and in Laxmipur it is usual to grow *arhar* (*kandula*). The legumes yield grain and rest of the plant is used as fuel for the family. The pod cover is used as feed for the animals. In turn, this leguminous plant enriches the soil and enhances the fertility. This is an age old practice of the tribes. The general crops under shift cultivation are:

CEREALS AND MILLETS

- *Echinochola frumentaca* (suana)
- *Penicum miliaceum* (common millet, suana)
- *Penicum italica or Setaria italica* (kangu)
- *Sorghum vulgare* (jona)
- *Pennisetum typhoides* (bajara)
- *Eleusine coracana* (mandia or ragi)

Plate 118: Steep hill cultivation with terrace, Kashipur

PULSE CROPS

- *Cajanus cajan* (*kandula* or *arhar*)
- *Dolichos biflorus* (jhota)
- *Vigra umbellata* (Dongrani)

OIL SEEDS CROPS

- *Guizotia abyssinica* (*alsi*)
- *Ricinus communis* (castor or joda)

Minor millets as ragi, suana, kangu and jowar, and pulses like kandula, biri, jhota (bailo) and oil seeds (*alsi*) are grown. The crop for the first year is ragi, which may be grown with black gram, *kandula* and jowar. The crop for the second year is suana, which may be grown with other crops like jowar and *kandula*. Suana is the major crop here. In the third year only *alsi* is cultivated. In most cases it is Dongrani legume creepers, where leaves contribute to soil fertility.

Depending on the gradient of the hill slopes, generally, the use and plough systems of cultivation are followed. Usually, it is the former. The choice of the land to be cultivated depends on factors, such as soil

Plate 119: Very steep gradient hill cultivation, Kashipur

fertility, manpower available to clean the vegetation, the amount of weed growth, the location and source of water. The shift cultivation system is based on:

(i) Choice of crops and duration variability within the specie
(ii) Organization of mixed cropping
(iii) The arrangement of short, middle and long-term fallow

The decisive factor in balancing these requirements is an understanding of the environment suitable to the soils and plants, piece of land and suitability to one type of vegetation or the other.

Mixed Cropping

The adaptability of shifting systems is closely related to three principles of cultivation, which are linked in various ways. These are mixed cropping, phased planting and crop rotations. This means the simultaneous growth of two or more intermingled, useful and economic plants on the same plot, which is an important characteristic of most shifting systems. The mixed cropping system is like more intensive farming systems. Generally, a few days after sowing the seeds of a cereal crop, they scatter the seeds of *Cajanus cajan* (*kandula*) on the same plot, by hand.

To a great extent, the choice of crop is dictated by the type of land, its shape and the angle of the slope. Pengos and Kondhs grow suana on must steep hills in the Kashipur block, and not *kandula* mixture. This decision is based on the root growth of the plant, its penetration into the soil, its hold and the time it takes to reach maturity. The undulated *Podu* land is preferred for ragi and *kandula* mixture. The most stony lands in Podu are best utilized by growing jowar, bajra and castor plants.

Physical-technical Reasons

Mixed cropping allows a fuller use of light, nutrients and water. Most crops have widely differing leaf canopies, which is effectively used for light. Nitrogen uptake is higher. The degree of root interpenetration seems to increase with a degree of crop dissimilarity, as different crops use different soil layers. There is therefore less run off and leaching.

Mixed cropping tends to reduce adverse conditions in the ecosystem. Disease and insect damage may be less and is likely to be more evenly distributed. The alsi around the border of mixed cropping deters entry of livestock.

There is a symbiotic effect between legumes and cereals under the system, which reflects in higher return per year. The increase in total output probably the most important reason why farmers prefer mixed cropping.

The Sabaras on the hills, have successfully utilized Podu cultivation. They have adopted about 25 to 35 species of plants under mixed cropping.

Socio-economic Reasons

This requires more labour per hectare than the other systems because it requires weeding and the vegetation to be cleared. The farmer may reduce risks by having several plots planted at different times but, under the condition of fallow systems, mixed cropping is the tradition and perhaps the more effective way of reducing risks.

In small hold farming, a varied food supply is usually sought. Over and above this, the householder relies on a continuous supply of fresh food, thereby reducing the storage. In this case the haryest of suana, a millet, starts from mid-August, followed by other millets and legumes like ragi, and *kandula*, etc.

Rotations

Depending upon the fertility of the soil, farmers take care of rotational basis of cropping system, the humus and nutrient content of the soil to regenerate gradually. Owing to the increasing demands of a growing population, the tribals are unable to provide an adequate rest period to their plots of land. This has a detrimental effects on the fertility of the soil and, therefore, production. This is a cause for worry as there is no fertile soil available in the forests. This is true of almost all hill blocks. Leaving the land fallow allows it to regenerate appreciably due to various factors, such as vegetation and regrowth, etc.

Characteristics of Fertilizer Economy

The fallow system appreciably regenerates the soil fertility due to vegetation, regrowth and several other factors.

Fertilizer Application in Shifting System

The system is associated with low degree of commercialization for sale of crops. In this type of farming, instead of using fertilizers, tribals rely on the natural fertility of the soil. The ash obtained from burning vegetation is the only fertilizer they use. In the case of certain root crops, like sweet potatoes, they use manure from their households. The green manuring or leguminous fodder crops, which can yield some material to animals and also enrich the soil during the fallow period can be one alternative. The introduction of this technology has scope and benefit. Traditional *Podu* cultivators like Sabaras and Kondhs are now using fertilizers, although they are only using urea for their rice crop. Their knowledge about fertilizers is very limited.

Characteristic of Labour Economy

The following labour inputs are required in shift cultivation:

(i) *Clearance of wild vegetation*: This is a most strenuous activity, requiring high labour input.

(ii) *Land preparation*: Immediately after burning, the soil is loosened on the surface and seeds are scattered at specific points, particularly the kandula seeds.

(iii) *Weeding*: The fallow systems suppress the growth of weeds and after the fire clearance, the soil is often weed free. But when the regeneration starts, in between the stumps, they are removed by the family periodically.

Scaring away birds from the crops is an important activity, whose responsibility usually falls to children, when they are on their way to school, and the older members of the family.

(iv) Harvest, transport of harvest and processing: This is a demanding task. As there are no roads or carts, the men in the family have to shoulder the load, literally, while women and children carry loads on their heads.

There is a particular pattern of division of labour the family members regarding bush clearance, stacking, firing, hoeing, seed sowing, weeding, watch and ward and harvesting. Responsibilities are also gender-specific.

Trend

As observed in a few progressive villages, trend is now moving towards sedentary farming. They now have terraced plots over hill slopes well packed with rubble and growing crops. The soil fertility particularly on hill slope is well managed. They are also knowledgeable about using the plough across the slopes, having perhaps acquired this skill from the Saoras, who can be said to be the best among all Koraput tribes. The Dongria Kondhs, however, have ramified to horticultural base utilization of hill lands, and are being suitably supported by agencies and are doing well.

There is a new trend of effective utilization of land resources through terracing, suitable cropping patterns and use of fertilizers in Laxmipur block whereas in Koraput block, Pottangi, Nandapur and Jolaput, there is a declining trend to this effect and this is a matter of serious concern for the inhabitants.

TABLE 8: FOOD PRODUCTION AND OCCUPATION TO KONDH FAMILY DURING A YEAR

Month	Men	Women	Children	Seasonal food
Jan.-Feb. Cold	All families individually thrash their harvested crops on threshing floors near to their hill-plots. Re-thach one side of roof annually and annual house repairs		Usually younger children are cared for at home	Beans, millets, maize and lentils
Feb.-March	Rotational 2 to 3 years fallow, new hill-plots cut and cleared. Cut the trees, pile, move rocks, dry and burn scrub and spread ash.	Cut and clear bushes and roots, burn them when dried. Do more hoeing than the men	Assist their parents in hill work, carrying food to family members on the hill.	As above, the food grain stock now delicing. More bevearages prepared out of grains. Sago palm juice favourite drink.
Mid-Mar.-Apr. Hot	Chaito hunting, preparing the hill plots. Collection of tamarind fruits	Tamarind seed separation, processing and sale	As above	Green mangoes (less millet, maize, etc., available). Beverages continue, Sago palm juice drinks.

(Contd.)

Apr.-May Hot	Weeding the podu plots	Collecting wild fruits and roots	Begin collective herding of goats and	Ripe mangoes, jackfruit, anything found in forest forms food.
		cattle. Others	help parents	
May-June Hot Thunder storms	Sowing seeds	Collection of ripe wild mangoes; collection of juice and dried mango stone collection, breaking of stones, processing for gruel at house	Tend cattle, goats. Collect wild mangoes. Help parents. Older children look after little ones	End of mango season, mango juice and pulp collection, stones, washed in water and dried. Mango juice dried for preservation after boiling.
June-July Monsoon rain	Looking after the cropland and interculture.	Collection of mango continues; collection of tolo seeds, where available.	Chase birds from ripe eafly crops of millets	Palm tree pulp or tamarind seed cake. Previous season's mango kernels powdered.
July-Aug. Rain	As in the previous month, and harvest small millets (both)	Wild tubers and fruit collection	Chase birds from ripening crops. Tend	Food scarcity, mango kernel gruel, tamarind seed gruel or sago palm
	Harvest bamboo shoots	Cattle and goats.	gruel or bread. Collect wild edible leaves,	Any available edible roots, mush-leaves, berries and/or rooms, etc. mushrooms and bamboo shoots.

(Contd.)

TABLE 8 (*Contd.*)

Month	Men	Women	Children	Seasonal food
Aug.-Sep. Rain	Harvest early millets with knife		Assist family	Naturally occurring leaf vegetables, quick growing varieties of cucumber.
	Pile up the cut millet on the plot		Tend cattle. Help parents in harvesting crops.	
	Harvest bamboo shoots	Bamboo shoot collection		Mushrooms finish. Earliest type of maize begins to appear. Bamboo shoot continue.
Sept.-Oct. Rain	As in the previous month, ploughing and *alsi* sowing. Harvesting of medium length millets, upland rice also.		As above	Root-vegetables and varieties of beans, gourds and wild forest roots, and newly harvested millets.
Oct.-Nov. Cool.	Farming together, harvest millet and beans with their knives (no sickles used).	Carry to nearby threshing floor	As above	Millets, beans, maize, rice, vegetables, Early morning snacks of millet gruel. Food plentiful and well

				mixed from mid-October to end of January.
Nov.-Dec. Cold, dry	Family continue to harvest millets, beans and now maize and rice.	*Alsi* harvesting start	As above appear.	Beans, millets, rice, ragi, and maize lentils
Dec.-Jan. Very cold nights.	Family harvests beans, lentils and turmeric	*Alsi* harvesting and sale. Kandula harvesting.	As above	Beans, millets, maize, lentils, rice and ragi.

Intensive Agriculture by Kolnara Kondhs and Malis

The Kondhs of Kolnara block of Rayagada as well as Malis of Koraput show highly skilled, intensive agriculture practices. Malis excel in vegetable cultivation and Kondhs in multicrop system. Their agriculture skills particularly the soil fertility management and moisture maintenance under the traditional system set up is well taken through generations

Malis of Koraput

The Malis are from the backward class. Their village is located at the foot of hills. Along the slopes lie their strips of cultivable plots, 15-25 ft. in length and 6-10 ft. wide, which have adequate supply of water from a perennial source. With balanced distribution, water used carefully. Among them, there is division of labour and women participate equally in work. They are very skilled at preparing their land, which is maintained free of weeds. They basically grow vegetables, growing multiple crops with varying periods of harvest. They grow between 3-5 species of plants simultaneously, on one plot. They start growing vegetables from Dussehra (September-October). After Dussehra, they sow raddish, French beans and amaranthus (kosala sag) together. During Diwali, they place ranja for support to the creeping French beans. They harvest the kosala sag leaves and raddish first, and then the French beans. This is how the cycle of different vegetable crops follows throughout the year. The highly skilled Malis follow a complex form of agriculture and depend a lot on compost.

The other crops that they grow at the same time include cereals like ragi, rice and even wheat during winter. These are usually for consumption at home and are grown in small patches on the edge of the fields and in plots adjacent to the road. The other crops that are of interest to Malis are maize, sweet potatoes, and sugar cane.

Plate 120: Multi vegetable species in single field

Plate 121: Off season cabbage-Mali, Nandapur

Plate 122: Raddish, maize, french bean in mixed cultivation

Plate 123: Mali women selling vegetables in weekly market

Nowadays, they have even cassava. The banana finds place on borders. Other vegetable crops are okra, brinjal, creeper beans, cabbages, and arrowroot, etc. They are equally good at floriculture.

It is taboo for Malis to keep goat and poultry, which can cause them to be declared outcastes by the community. They keep sheep instead, as the community depends on horticulture for its livelihood.

Kondhs of Kolnara Block, Rayagada

The Kondhs of Kolnara block show great skill and wisdom in growing food for the family. All, including children, stay on the family land through out the day and the men continue to guard the crop during night. They have longitudinal plots. They grow pure strands of ragi, sorghum, *arhar*, oil crop (sesame) and vegetables like brinjals, ladies finger, some creeper beans and chillies. Most of this is meant for consumption at home, although some of it is sold in the open market.

Their plots are wider than those of the Malis, and they have to carry loads of water on their shoulders to irrigate these fields. The family, including children, join in this endeavour, whereas among Malis children do not work in the field.

Each family plot has a *machan* or a platfrom, raised on four strong wooden pillars and permanently fixed in the field. During winter, a small fire is kept burning beneath it, which not only keeps the animals away but also provides some warmth. Children rest here during the daytime, while the other family members gather here for lunch and to rest. As mentioned earlier, men spend the nights here, guarding the field.

Their land preparation is excellent. The fields are well levelled, and the soil is well pulverized and soft. Their plots are maintained clean and free of weeds. They use cattle to prepare their plots.

The most distinctive feature here is the intensive agriculture, under rainfed conditions. They have traditionally selected crops, which have minimum water requirement. What is most interesting is that they carry loads of water on their heads, to irrigate their fields during the *rabi* season. They also follow transplantation of ragi and sowing crops in a line.

The other feature in this agriculture is the use of fresh cow-dung and urine. Cultivation takes place on a rotational basis, on longitudinal strips of land. This shape is advantageous as it allows bullocks to enter

Plate 124: Night calf pen and night watch tower in the field

Plate 125: Evening watering of vegetable plants

Plate 126: Fixing pegs for cattle over the plot for urine and dung

the plot for land preparation, while the crop is till standing.

Most families have movable calf pens, made out of bamboo, which can hold 5 to 8 calves. Calves are separated from mothers during night to get morning milk from the cows. In the afternoon, men fix cattle pegs in the fields and tie them on return from grazing. The night urine and cow-dung enriches the soil, and field yields more crops. Buffaloes are tied little away from standing crops and close to the periphery. They follow the stall feeding system for management of bullocks.

The Kondhs of Kolnara block of Rayagada, as mentioned above, give the greatest priority to ragi, of which they have two varieties. The early variety is harvested between September-October, while the late variety is harvested during November-December. Ragi is the staple food; they cultivate it in about 30 per cent of the family landholdings. The late variety of ragi is grown alone, while the early variety is always grown with bailo, a type of local beans. There are two varieties of bailo. One is suitable for hills, and stony land, while the other is suitable for good soil and cultivable land.

Bailo is sown first, followed by ragi, which is transplanted as a part of the mixed cropping system. After the ragi is harvested, bailo takes over and flowers by December.

Most of the land holding, up to 40 to 60 per cent, is covered by a local variety of *arhar*. It usually grows to 6 ft in height, with some even reaching up to 7-8 ft. *Arhar* is generally grown as a solo crop but some people like to grow sorghum with it, which matures for harvest during December-January. Some farmers place few castor plants in the crop.

The land used for growing *arhar* is used to grow ragi in the following season, and vice versa. Unlike the other Kondhs of the hills, Kolnara Kondhs claim that growing ragi as a solo crop is profitable. These Kondhs do not follow the system of mixing the ragi and *arhar* crops.

The next most loved crop is tobaccos, which is grown for personal consumption. All adult males and females smoke cheroots. After harvesting ragi, they transplant tobacco plants in well fertilized and clear land. The crop is planted and grown in lines and is watered for a month. After three months, the crop is ready to be harvested. Tabacco is grown in winter.

In October, families grow some tomatoes, chillies and local brinjals, the latter being planted in a few plots. The tomato and brinjal plants can get by without too much water. Marigolds are planted on the edges of the tomato plots. Horsegram is also grown in small plots. While the winter rain is harmful for the bailo crop, it helps the horsegram and the *arhar* crops. In this way, the advantages and disadvantages balance each other out.

Around these plots they grow the local, semi-beans creeper called 'jhata'. This fulfils the domestic requirement of vegetables during winter.

Some farmers cultivate small plots of cotton. Others, who have land resources, also grow paddy.

There is another oil seed crop sesame, which does well in conditions of low soil moisture. It does well in winter. About 10 per cent of the land is used for this crop alone.

Wherever available on the hills, cultivable land is used to grow sorghum, pearl millet and *arhar*, etc. Their farming system is intensive and strikes a good balance between crops grown to meet domestic needs and those grown as cash crops. Families labour collectively on the fields, up to February-March and then again from May-end onwards.

Biological Control in Tribal Farming System

De Bach (1964) defined biological control as 'the action of parasites, predators, and pathogens maintaining another organism's density at a level lower than would occur in their absence'. This has been operating on plant resistance, so as to make the environment less favourable to the pest and more favourable to the pests' natural enemies. The subject has been well detailed out by Panda and Khush (1995). The plant resistance with cultural control has been studied as herbivore phenologies, vegetation diversity, variety mixtures and trapping.

The herbivore phenologies, the classical example is an effective control of bore weevil and pink bollworm through reducing the over wintering insect population has been developed. Similarly, in Koraput, the tribals, Kondhs have the same over wintering insect population and saving the pod formation from cloudy weathers has been devised through genotypic selection of *Cajulus cajunus* long duration varieties, thus enabling it to produce without use of insecticides.

The vegetation polyculture, is where more than one crop is grown simultaneously in the same area. It is one way to increase diversity. This culture practice is ecological complex because interspecific and intraspecific plant competition occurs simultaneously with herbivores, insect predation and insect parasitism (Van Emden, 1965). The elimination of alternate habitats may lead to decrease predator and parasite population and thus add to increased insect pest population (Southwood, 1975). The *kharif* season crops with vegetation diversity in tribal farming systems particularly in Sabara hill cultivation is a classical example of polyculture.

The variety mixture concept, has the unique feature of polyphagous insects where the feeding preference is altered by including genetically different crop cultivation in cropping area. This is the common feature among the hill cultivators. As already described, this feature provides a continuous supply of food to the family after July and August till March, and the mixture of legumes with cereals adds to the soil fertility. The system is common among Sabara hill tribals, Bondo hill tribes and Kondhs and many others. The former Myrada rice cultivation among Gutob Gadabas was one such system. The variety mixture was small millets. It exists among Koyas.

Plant strands that are grown to attack insects and other organisms, so that the target crops can be saved from pest attacks, are known as trap crops. The target crops are protected by either preventing the pests from reaching the crop or by concentration them in a certain part of the field, where they can be destroyed. Cotton, soyabean, sesame as well *alsi* and sorghum in ragi crops are examples of trap crops in the tribal farming system.

The trap crop system mechanism works in standing crop through diverse plant population. A specific insect is less likely to find its host because of the presence of confusing and masking chemical stimuli, physical barriers to movement and other adverse environmental factors. This lowers the insect survival rate (Panda and Khush, 1995).

Frog population is usually a good indicator of the effectiveness of a bio-control programme in paddy fields. It provides an indication of the use of pesticides. Heavy usage eliminates snakes from paddy fields, as also crabs and frogs. The Gadaba farmers give a promise payment and prayer to these animals. It runs:

Oh! Jolokamini, Patal-Kamini; Oh! Bund living people, guarding people; Oh! Watch and ward people, we have brought for you food and gifts. Jolokamini is the goddess of water, Patalkamini is goddess of underground, the Bund people are snakes, periphery and watch and ward people are bears and wild animals, like tigers and leopards, etc.

Three decades ago in the Koraput district, tribal farmers used spiders to control insects from causing damage to their paddy fields. The practice is still being followed in the neighbouring south Bastar district. Wolf spider (*Lycosa pseudoannuclata*) is used for the purpose. This method is totally lost from Koraput tribals agriculture. Spiders' nets on the crop, which is spread all over the field, collectively give an impression of a fine, transparent sheet placed over it. This is particularly true when the dew drops on the crop catch the early morning sun and glisten. The tribal farmers take out the spiders from the fields before harvesting paddy and place them on mango or other suitable trees where they live safely for the rest of the year. When the crop needs protection from the pest attack again, they take out the spiders and place them over the field. Spiders are kept at an appropriate distance from one another, to allow each to spin a web, as described earlier. Of course, much work has been carried out and reports are available in this sphere, particularly the spider base bio-control.

In some places, tribal farmers still use a wild growing local creeper, Guhamala, which means 'night soil creeper'. It has a very pungent smell and works as an insecticide. It is laid over the standing water in a paddy field that has been infested with pests.

A weed plant Kolthia (*Tephrocea purpurea*) is known to control paddy caseworms (Parasar, 1992).

In the Jeypore-Koraput valley, tribal farmers erect a branch of each of the following plants during Bandapan: *Kokributa, Bonobhalia patro*, *Batri patro*, *Bounso danga* and *Kendu danga*. I am not sure how it works but this ceremony is performed on a fixed date, and is performed to prevent the insects, or 'chitta' from attacking the paddy corp. The Matia and Koya tribe in Padmagiri area of Malkangiri use the leaves of bitter gourd, the distilled sediments from *mahua* flowers, *Bonobhalia seeds*, and the cuticles of *machi* tree as pest control devices.

The tribal farmers use jackfruit leaf extract which is sprinkled on newly sown *alsi* fields to check *cascuta* weeds (Dixit, 2001, personal communication).

The tribal farmers plant *alsi* (*Guzitia abyssinnia*) around the border of millet crops on uplands of Koraput district to deter the entry of livestock into the fields. The smell of the *alsi* plant repels the livestock and has a deterrent effect on pests as well. Studies in Israel have shown that 'phermones' from the feces of lions were used to keep away deer and antelopes. In the field of Rayagada Kondh, marigolds are planted around and in between rows of brinjal, to protect it from the wilting disease that is caused due to nematodes. The mixed cropping system followed by the hill tribes needs to be studied, to gauge the effectiveness of their disease and pest control methods.

The white ant population is kept in check in the fields, woods and forests by the bears, who are very fond of them. In villages, poultry keep the white ants in check-up to the point they can reach. The African experience of sprinkling cattle urine over thatched roofs, to protect them from white ants is not known here.

Plate 127: The ergets feeding on ectoparasites of cattle—Mathili

Ticks are most harmful to the health of the livestock. They suck the animals blood, cause irritation to the skin and also transmit many blood parasite diseases, like anaplasmosis, theileriasis and babesiosis, etc. Egrets help to control the tick infestation in the livestock. The egrets become active over the grazing cattle and buffaloes from September-October and pick out ticks. They continue doing so till the end of summer.

The Crop and Livestock Losses due to Wild Animals and Accidents

As most cultivated fields are adjacent to the forest, wild animals are a constant threat to the tribal farmers, often contributing to the loss of crops.

The wild animals that usually damage crops are wild pigs and antelopes. In the Rayagada area, elephants do most of the damage. If not controlled properly, pigs and livestock from the village can also cause sever damage to the crops. The tribals dread this because the loss of a crop causes them, and their families, great suffering. After eating crops like paddy and tubers in home gardens, wild boars cause further serious damage by rolling on the crop. These animals, wild boars with elephants, are the potentially limiting factors in the popularization of tuber crop agriculture, like yams and the newly inducted tapioca. This was witnessed in the Phulbani district and Bissam cuttack Kondh villages.

Wild birds damage crops like small millets, pearl millets and khed jona (sorghum). They do most damage at sunrise and sunset, during the grain formation stage and during the harvesting period. Birds damage prospects of good crops by eating the germinated paddy in the fields, specially the sprouted sown seeds.

Bats are a threat to fruits in some areas. With a dwindling bat population, however, this threat has reduced greatly. The belief that the flesh and blood of bats has great medicinal value, especially for those suffering from asthma, has had a direct bearing on their population.

Wild field rats also damage standing crops, as well as vegetables and bananas in the backyard farms. Their numbers are kept in check by snakes. The use of heavy pesticides these days, however, has had a detrimental effect on the snake population, allowing the rats to multiply uncontrolled. In the 1970s, the Malkangiri region was dreaded for its posionous snakes, but not so in 2001. During summer, young tribal boys remain busy searching for rats, which they then roast and eat. They love eating rats and it is a common sight to find damaged bunds in paddy fields, which have been dug out in search of rats. At times the tribals kill and bury rats in their fields, hoping that the rotten smell that follows would deter other rats from entering.

The loss of crops to wild animals is a serious problem and requires drastic steps to check it. Tribal men, therefore, who are usually able-bodied, spend nights in their fields to keep wild animals and thieves at bay. For this, they build raised platforms on four strong pillars at, or above, 6 ft and make thatched roof over it. Staying on the platform usually keeps them safe from most wild animals, except elephants. In Laxmipur and Kashipur areas, Kondhs make an intelligent device out of bamboo which, if pulled once, makes noise for several minutes and is enough to scare away the approaching animal. In some places they communicate with others, on similar duties in neighbouring fields, either by shouting or through whistles made out of pipes. This parctice is popular with the Koyas. The Bondo make special devices to scare birds away from their grains. It is considered a taboo for women to do this job and spend nights on the platform.

The Bhattras prepare a special device to scare the birds. It is made out of two dried crab shells, which are tied to a dried leaf with a string. This in turn, is tied to a stick. When the leaf blows in the wind, the shells make a cracking sound. In the Kashipur hills, the Pengos also make several devices to prevent birds and animals from entering their crop fields. Some of these operate automatically in the wind, while others are tied to strings and pulled from the night shelter. All these devices make loud sounds to scare away the animals.

This crop watch duty continues from September to December for regular crops, and can be extended up to February and March if *arhar* crops is in the field.

Accidents happen from time to time, when tribals are charged by wild boars, bitten by snakes, attacked by bears and sometimes even by tigers and leopards, etc. Bear attacks usually take palce in villages where *mahua* trees are common. These attacks usually happen early in the morning, when people go out to collect the fallen *mahua* flower and run into bears.

Plate 128: Night watch and ward shelter 'Mancha', Kondhs on hills

Plate 129: Wind propelled dry gourd bird scarer in small millet fields, Kashipur

Plate 130: Wind propelled in box animal scarer in small millet forest field, Kashipur

Among most tribals of the plains, killing bear is taboo. Bears are attracted to ripe jackfruit and the maize crop grown in the backyard plots. They also damage the banana crop, although the local banana plants are very tall and escape damage. Besides people also make pits close to it, which not only gather water during the rains but also keep the bears away. In most tribal villages, dried thorny plants are tied to the base of the jackfruit tree. This helps to keep the bear away and save the fruit.

Black faced langurs and the general monkeys are found in the forests. They do not cause much harm to the tribals' crops. While the Sabaras kill monkeys, among many other tribes this is considered a taboo.

Pigs are a constant threat to the home gradens and, as mentioned earlier, in some villages, almost all the pigs are sold off at the oust of monsoon and re-purchased in November. In some villages, they retain a few pigs, which are kept in temporary enclosures, with limited food.

Up to 2 to 3 weeks after sowing, the backyard crops are under threat from the village poultry, with ducks being the prime culprits.

Traditionally, in Malkangiri and the hill regions of Koraput, the livestock are left free through the day. This parctice is followed from December to the end of May. The situation demands fencing around the crop fields.

In the season when *mahua* flowers are collected, the Koyas do not release their livestock for grazing till 10 to 11 a.m.

Various plants are used as live fencing around their fields, especially their backyard farms, to protect their crops from livestock and other animals. However, this requires further study and documentation.

Having good knowledge about the probable pregnancy periods of wild animals, which are seasonal in nature, tribals avoid hunting or killing them at this time. For, some reason, the Bhumias do not eat fish during the month of Kartik, perhaps because pregnancy among animals is common at this time. The preferred time for hunting is April and, if they have a choice, they would rather hunt the male animals first.

Food and Crop Losses from Wild Fire

Fire is major hazard that the straw-thached houses of the tribals, such as Kondhs, Sabaras and many others, face. Built close to each other, across very narrow streets their houses are very vulnerable to fire and can lead to total destruction of the village. In such cases, the poultry, pigs and household materials, seeds, conserved food and cash, clothes, beds everything is lost. I witnessed such devastation in one of the villages of Laxmipur block.

The top priority for most households, therefore, is to have a tiled roof, which they get as soon as they are able to afford it.

Forest fires during summer also cause untold damage, and have a devastating effect on food production in their primarily forest-based economy.

Forest Predators—Livestock and Poultry

Predators like tigers, leopards, jackals, groups of wild dogs and foxes, etc., constitute some of the most unwelcome visitors to tribal villages from the neighbouring forests. They carry away dogs, pigs and poultry from villages, sometimes even human beings and, from the grazing herds, sheep, goat, cattle and buffaloes. The small population of sheep in the Laxmipur block can be attributed to predators. In 1995, the following survey was conducted in 100 villages, about the predator menace that the tribals' livestock faced:

TABLE 9: NUMBER OF VILLAGES IN THE FOLLOWING BLOCKS REPORTED PREDATOR PROBLEM

	Jeypore	Koraput	Kundra	Laxmipur	Total %
Tigers	06	04	04	06	20
Leopards	06	15	07	15	43
Jackals	05	10	01	03	19

The wild dog menance is reported from Laxmipur block, among others, while poultry poaching by foxes is a common occurrence. In villages adjacent to forests, predators account for the limited growth of the livestock population. Due to the construction of new hydro-electricity dams i.e. Kolab, Indrabati and Balimela, attacks by wild animals are concentrated in some pockets, leading to heavy losses of livestock in those areas. This needs serious thinking and prompt action.

The white ant population is kept in check in the fields, woods and forests by the bears, who are very fond of them. In villages, poultry keep the white ants in check-up to the point they can reach. The African experience of sprinkling cattle urine over thatched roofs, to protect them from white ants is not known here.

The Agriculture Systems: Transition, Adoption Trend

During the lean season, when food is at a premium the forest helps the tribals tide over the shortage with a fairly regular supply of food. This is usually in the form of edible roots and fruits. Owing to the increasing demands of a growing population, the forest is among the first casualties of subsistence agriculture and its pursuit of enhanced food production. This increase in the number of households either leads to extension *in situ* or a migration into other forest areas and, thus, formation of new villages. The other alternative in this situation is the extension of cultivable land through the slash and burn method. This is successful with multiple-specie crops and intercultivation with legumes only. This method is driven by the fertility of the soil.

In subsistence agriculture, there is a gradual reduction in the spectrum of cultivable crops. The aim is to produce marketable goods, as well as food for consumption at home, from cereals. At this stage, legumes are replaced and cereal crops dominate the situation. Many varieties of less but valuable food producing crop are thus gradually lost from the system. Many view it as major loss of nutrition for the families.

This situation leads to a semi-intensive crop production system. It demands high production from productive soil initially. The system adds up fertilizers, and is followed by improved varieties of crops. The pesticides appear at a later stage, when losses are constantly building up in the system. As mentioned earlier, the system builds up rapidly due to increased demand from the families. Of course, the quality of production depends on various other factors.

Food production skills are introduced into the system on a priority basis. Primarily, the factors contributing to the loss in production are taken care of. The first of this is the removal of weeds, which depends on labour. By the time collective family labour is available, unfortunately, weeds have usually grown and inflicted losses by absorbing all the nutrients from the soil.

The first generation of tribal farmers do not like to try the new improved crop varieties, of even the improved high yielding traditional crops. They fear that it may lead to a total failure of the crop and deprive their families of food. In the meantime, some non-tribals around the area take the plunge and introduce these new, improved, high yielding varieties of crops. Some are successful, some are not. This continues for 5-6 years, during which time the tribal farmers keep an eye on their neighbours' fields. If they find that the new techniques have clicked, some of them adapt it for their own fields. This is true of the paddy crop. In the last three decades, less than 30 per cent of tribal farmers have taken to improved paddy or other, seeds.

The adoption of chemical fertilizers is a gradual and slow step. Absence of training in their use means that chemical fertilizers are slow to find their way into the tribal farming culture, while not many are aware of the benefits of chemical fertilizers, they are familiar with urea because they have seen the impact it has on the growth of a plant. They believe that urea is all that a plant needs to grow. Most tribals are unaware of phosphate and potash fertilizers. The hill tribes believe that chemical fertilizers harm the land, have a detrimental effect on its fertility and make it dependent on them. A scientific study must be conducted to verify this claim. But since fertilizers have been used on only 10 to 20 per cent of the land here, the above claim can hardly be said to be true to the district. However, the possibility that such situation can appear under acidic pH soil as on hills of Koraput cannot be ruled out. And if composite NPK combination is used, with required doses, along with organic manure it could show the way for further improvement of crop production.

There is another problem with in this technology. The application of nitrogen to the soil, its assimilation and production of crops is also dependent on the genotype. Most of the indigenous varieties of crops respond to nitrogen. They show vegetative growth and lodging problems, but not in yield. So the fertilizer use should concurrently be related to genotypes responsive variety selection too. This is a drawback of chemical fertilizers.

The other important aspect is correction of acidic pH of soils, and the deficient micro and macro nutrients. These soil have both iron and aluminium with a high toxic level. The phosphorous deficiency is another major limitation. Boron and zinc deficiencies are apparent from studies as well as the experiments with corrective methods of yield. This intervention has its limitation from economic point of view, availability of the chemicals, knowledge of farmers and utility skills. For the tribals engaged in productive agriculture, perhaps this can be one of the schemes that the government can provide aid to in the future.

The Tribal Agriculture in Transition

Under traditional agriculture, among tribal societies, the food production system from agricultural crop and animal husbandry can be termed sustainable. In the present situation, more needs to be produced, to meet the demands of a rapidly growing tribal population. In other words, the sustainable food production system has to evolve into a new farming system, which will also have to be made sustainable. It is important and vital to work carefully under poverty situation in traditional tribal farming societies where logics to produce food is in built from centuries knowledge. It is most essential that during transition phase, this logic in food production systems need to move *in situ* in the process to attain new improved production goals.

The challenge is a food production system development plan, which generally ends up at stratification of crops and species of livestock for high unit production. Further, traditional subsistence farming system will progress in phases from family food production to market oriented production economy. The traditional tribal farming is very vulnerable here. Because, throughout the stages of development which it will go through various constraints like input supply, inappropriate technology, lack of proper support at credit accessibility, market and marketing skills. These stages require lot of other social and managerial skills support in order to be successful.

Therefore, it can be said that, in a poverty-stricken society specially among primitive tribal groups, early maturing crops are vital in the food production system. This takes care of hunger during the most difficult periods of food shortage at home. The objective here should be to improve the production of these traditional crops.

There is a steady but gradual food production supply line. It is functional and needs to be more efficient and productive to meet future demand. Initially, the interventions may be through propagation of existing improved genotypes for higher production, as well as correction of factors, which decrease the production.

The large spectrum of plant species that are used in the food production system, still needs to be maintained because it will tend to shrink once the process of development gets into it. The farmer needs to be sensitized to this inbuilt, effective food system operating since centuries.

Early maturing livestock and prolifically breeding species are the most useful food producing components of the poverty-stricken tribal life. Most people own poultry. Domestic pigs are the next in line of valuable species.

Goat is another important animal, which is reared because it matures and grows quickly and breeds prolifically. So too do sheep, who are also resistant to many diseases that afflict goats. Among goats and sheep, there is a large variations in terms of early maturing characteristics and breeding habits.

Poultry, pigs, goat and sheep provide food to tribals. But, nowadays goat and sheep find handy market and are sold for cash to meet family needs.

Trees like mango, jackfruit and mahua provide food to the tribals during summer and early monsoon. They help them through the days of severe food shortage. Cash from sale of tamarind, and bamboo shoots, provides food for the family.

The backyard farming system, which has been described in detail, is completely under the charge of women. This system provides a regular and assured supply of fruits and vegetables for the family.

The trend is towards cultivation of improved and high yielding varieties. The shift has already taken place and it is evident in the 'maize' crop. There is not much known research or improved variety of seeds in creeper beans. This is an important crop that meets the vegetable needs and protein requirements of tribal families and further research on the subject will be a great help to the tribals of central India. The genotypes of pumpkins, ridge gourd, bottle gourds are very good, although they are endangered by the incoming genotypes from market seeds. The bitter gourds are now mostly the improved varieties; the little gourd, 'tundri' variety, will remain unchanged due to its economic benefit and taste factors. Some vegetables and fruits, like drumstick and papaya will be taken up in future with the improved varieties.

The more popular vegetable crops are tomato, brinjal, ladies fingers and the newly introduced French beans. For vegetables like cauliflower and cabbages, etc., seeds from the market are generally used.

Storage of good genotype, disease-free, high-yielding seeds is most important. Apart from beans, gourds, pumpkins and maize, farmers need advance knowledge about collection and storage of other vegetable seeds. This requires training.

Growing vegetables is not considered a gender specific task.

The ragi crop in recent years is showing a diminishing trend, but it is still the staple food for all the tribals. The research taken by ICAR and Universities of Karnataka and Tamil Nadu and the few varieties from OUAT will be of interest. In order to improve the existing ragi production level in tribal farming system, it is vital that the package of practice followed need to be more efficient and productive. It shall be of interest for extension practitioners in the future. The uplands in rainfed agriculture shall still depend on ragi and other small millets because upland rice cultivation has its limitation.

The small millets group will remain an important source of food, particularly in early rains when there is scarcity of food for all hill dwelling tribal groups. This has its immense value due to comparative gestation of crop, and land type and response to fertilizers, etc. Even though it is favourite with the poor hill tribes, not much research has been conducted into the subject. Being an important food for the poor, all research studies about the small millets groups should be coordinated.

Crop Stratification

There is a classical example of agricultural crop stratification in Malkangiri and Nabarangpur areas, Dandakaranya (DNK) Project area interventions, among the refugees of Bangladesh. The agricultural interventions started here in the early 1960s.

The project in initial years, specially in Malkangiri zone, targeted the skill and resource base of the refugees and adopted 'mestha', an inferior jute crop. This was grand success. It had a very good marketing and export system well built with the target group. This was extended to 110 villages and about 10,000 families, with at least 4 acres under cultivation by each family. In the 1980s, the declining use and, therefore, the market of 'mestha' forced the farmers to adopt *til* as the new crop. It was less labour intensive and more eco-friendly, reducing pollution of water.

The current favourite, however, is the groundnut crop, which has reduced the emphasis on *til*. The groundnut crop is preferred for its competitive price, its production potential and high yield.

Maize was introduced as the major crop by the DNK project in the Nabarangpur zone in the 1960s. The crop pattern remained unchanged for three decades and had a great impact. The quality has witnessed an improvement in the colour of grain, size and numbers of cobs and period of maturity. It can be said that the growing market and price structure, along with quality and reliable composite varieties for years, the recent hybrids, and the winter adoption maize have kept the process going. The package of practice for high yielding maize has been standardized and farmers are using the same gainfully.

The said development was on a regional basis, with a dynamic export marketing component in it.

Changes have also come about among the subsistence tribal farmers, like the Koyas. Low acceptance of these changes can be attributed to a host of complex reasons related to labour, economy, skill and marketing support. The most important factor is the situational shift from subsistence farming to a market oriented farming system.

Natural Stratified Agriculture among Tribes

The tribal groups have varied agricultural base. This can be seen from the following:

The Dongria Kondhs of Rayagada depend mainly on fruit trees, which bring in most of the cash and food for them. Jackfruit, pineapple, banana, mango and orange trees, along with those of seasonal ones, can be seen all around.

The Desia Kondhs are dependent on soil fertility and the crops. The common practice of human sacrifice, followed in the eighteenth century, is an indication of the importance that the tribe attaches to the fertility of soil. The Kondhs have many other ceremonies and rituals to propitiate various deities for an assured harvest from their fields, trees and forests, etc.

The Hill Sabaras still follow traditional hill cultivation of mixed species agriculture, as do Bondos and hill Didayis. Their efforts are orientated towards the production of small millets. All tribes that follow hill cultivation, or the slash and burn method of cultivation, depend on millets as their major crop, followed by legumes like *arhar* or *kulthi*, and oil crops like *alsi*.

Strikingly almost all Austro-Asiatic tribes, like Bondos, Sabaras, and Gadabas, have a peculiar affinity to rice cultivation. It may either be on the hill jholas, terraced lands or plains. Their cultivation method is still perfect.

Elwin (1964) wrote, 'The great achievement of the Saoras is their terracing. They build up the hill sides for a thousand feet, terrace rising above terrace, perfectly aligned, so carefully done that not a drop of water escapes uselessly and all erosion is checked. Sometimes a great stone wall, 15 ft high, well hold up a ridge of cultivable soil only 3 ft wide.' These terraces have been rightly praised as works of great engineering skill.

The tribals dwelling along riverbeds, like Kolab, Indravati rivers have the skills to cultivate crops like kusum, castor, brinjals, pumpkins and gourds under least water requirement condition. They have the skill to use the moisture of the soil to its optimum.

Characteristics of Animal Husbandry in Tribal Farming System

Animal Husbandry Systems

Animal husbandry is one of the most impoıtant components of tribal agriculture, and livestock is one of its main natural resource base.

Livestock raising is carried out in many forms, usually referred to as production systems or agro-ecosystems. The production systems have evolved as a result of the agro-ecological potential, the relative ability of land, labour, capital and the demand for livestock products. It may be at family, village or regional levels. Based on the degree of integration with crops and its relation to land, the world's livestock sector has been classified in three broad livestock production system (Sere and Steinfeld, 1996).

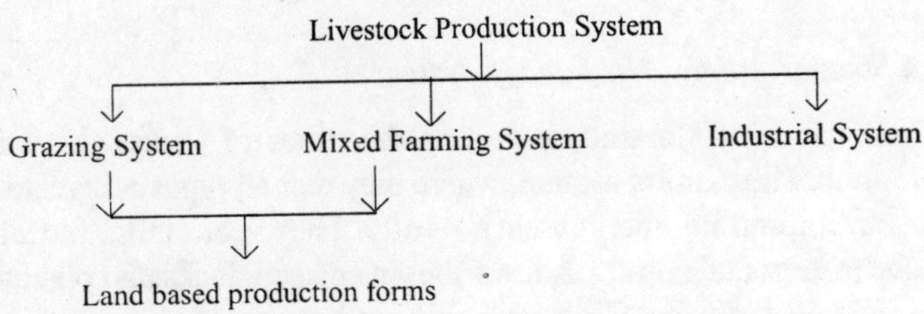

Grazing Systems

This system is exclusively about livestock production, with little or no integration with crops. It is mainly based on native grassland. The agro-ecological aspects strongly influence the nature and scope of livestock-environment interactions in grazing systems. The traditional tribal animal production system is mainly based on this system. Cattle, buffalo, sheep and goat husbandry come under this category. Grazing systems supply 30 per cent of the world's production of sheep and goat meat; in Koraput this system provides 100 per cent sheep and goat meat.

Mixed Farming Systems

In this system, the crops and livestock production are integrated on the same farm. It contributes most to the production of total milk and meat. Resource use in mixed farming is highly self-reliant as nutrients and energy flow from crops to livestock and back. This system appears under improved farming conditions due to the improved crop production system. It has irrigation facilities, improved varieties and a good functional marketing system. After the grazing system, the livestock production tend to follow into the mixed farming system.

Industrial Systems

These systems cover industrial types of production and small scale urban or peri-urban production in developing countries. Both monogastric (pig and poultry) and ruminant production systems exist. These systems are open both in physical and economic terms. They depend on outside supplies for feed, energy and other inputs. They are strongly market driven. The industrial system does not appear in the traditional farming system.

Mixed Species Animal Husbandry System

All these tribes of Koraput, including the tribes of Orissa, have an unique animal husbandry system, where they rear all types of livestock species in limited numbers in each family. They rear cattle, buffalo, goat, sheep, pigs and poultry. A few tribes even keep ducks and pigeons. For almost all the tribes, livestock and poultry serve the same basic purpose but every species has its own special importance, which probably depends on ecology and requirements of the concerned. They breed poultry for game and mutton, but not for egg. Poultry is an important component in religious services and entertaining visiting family guests and friends. They breed cattle and buffaloes primarily for manure and draught power but some pre-farming tribal societies rear them mainly for their food value. They prefer cattle to buffaloes because the former breed more prolifically. They utilize both male and female cattle for agriculture work. They show preference of cattle over buffaloes for prolificacy. Most of these tribals do not milk cows and, in fact, barring a few, they are not even aware of the uses of milk. They breed sheep, goat and pig for food value. They mostly prefer goat to sheep from commercial point of view. Though they claim that sheep are having higher disease resistance and goats more susceptibility to diseases but still prefer goats. They love pigs for the profits these animals bring in. They are very fond of pork but hate the pig for the havoc it can cause in their fields and backyard farms. Sheep, goats, cattle and buffaloes, on the other hand, graze in the wastelands or in the forests around the villages, and require no input from tribal families for their upkeep. Pigs are scavengers, living on the water used to wash the utensils and rice bran. The poultry live mostly on organic materials or insects around the house or village. Only roosters and newly hatched chicks receive special attention. The tribals do not use cow-dung for fuel, but preserve it carefully for fertilizing the land. They collect sheep, goat, cattle and buffalo dung to use as manure.

The energy and economic efficiencies of different animal husbandry systems, such as poultry, swine, goat, cattle and mithun of the Nighis, Kharbis, Kancharis and Chakmas of Arunachal Pradesh in north-eastern India have been compared by Ramakrishnan and Gangwar (1989). They have observed that all except Kharbis practice swine husbandry. Kharbis and Chakmas rear cattle for milk, while the others do it for the meat. The energy and economic efficiencies of the mithun and the cattle raised for meat are reasonably high. Poultry and swine husbandry are largely village based and are relatively less dependent upon forest, unlike cattle and mithun that are sustainable only when the forest resources are high. The differences in animal husbandry are partly due to cultural differences among the communities. The Nighis have better developed animal husbandry and consume more food than others through this sub-system, although, on a per capita basis, the export earning of the Chakmas is higher. Poultry and swine husbandry are the two animal husbandry systems in the village. Both are largely detritus based among the Khasis in Meghalaya. The Kondhs, Bondos, Koyas, Gutob Gadabas, Sano Poraja, Bodo Poraja and the Scheduled Caste rear pigs; Bhumia, Omnatyas and Bhattra, however, do not keep pigs at home. Pig rearing is, thus tribe specific. However, the Dongria

Kondhs, Gutob Gadabas, Kondhs have specific interest in buffaloes. Almost all the hill tribes have a particular affinity and liking for animal flesh. This has been observed in the Bondos, amongst the Dongria Kondhs and all other tribes. The cash that they earn either goes towards fetching a wife or towards buying a beast for the feast. The Dongrias are extremely fond of buffalo flesh, and are even known as buffalo-eater-Kondhs by other local tribes. Bondos do not hesitate to attack the neighbour Didiais for forcefully stealing cattle from their village (Elwin, 1950).

Mishra and Ramakrishnan (1982) wrote, that the compost pit in a Khashi village is a very important component in the functioning of the eco-system. It provides feed for secondary production and generates organic manure of plant and animal origin, which is cycled back into agriculture. Animal husbandry formed an important link in the detrital food chain by utilizing the garbage and vegetable waste of the agriculture system. The forest, apart from providing the basis for agriculture, also meets part of the fuel requirement of the village.

The sheep, goats, cattle and buffaloes go out to graze in mixed herds, to village forests, wastelands and harvested stables. Pigs, poultry and ducks are their other animals of choice. This is classified as Mixed Species Animal Husbandry System. Almost all the tribes of Koraput district follow this system. This has originated and remained viable probably due to the risk of losses in epidemics. The benefits of modern medicine have still not reached the primitive tribals, especially the hill tribes; because of inaccessibility to their habitats and lack of facilities to preserve vaccines. The tribals feel secure in their traditional system and are able to sustain some loss of livestock. But when they begin to lose animals to some kind of epidemic, they usually describe the loss to an enemy or a wandering evil spirit, etc.

Only few tribes, like the Koyas of Malkangiri and Bhattras of Nabarangpur, have passed on from a Mixed Species Animal Husbandry System to the Specialized Animal Husbandry System. They keep cattle and goats, and least prefer buffaloes or sheep. So they graze flocks of goats or cattle separately. The species specialization is a reliable indicator to judge the level of tribal animal husbandry in a given situation. The number of livestock is an indication of the wealth and social status of the owner.

Tribal Women and Animal Husbandry System

The tribal women play a vital role in their Animal Husbandry System. They take total care of housing livestock and poultry, cleaning and collection of manure, arrangements for hatching eggs by brooder hens and care of the newly hatched chicks. They also feed pigs, and take care of piglets in the village. The kids and lambs remain under the direct supervision of the ladies until they leave for grazing. They keep a daily count of the household livestock and poultry and inform the male members of the house in case of any loss.

In some tribes, older women also accompany the younger boys for grazing the herd. A detailed account has been given in separate chapter.

Livestock Species, Household in Tribal System

Under permanent cropped high lands in Koraput, families keep few heads of cattle and high density is reported. A greater number of cattle are found among the people of the hill tribes than those living in the plains. However, buffaloes can be found in an equal measure in Jeypore, and irrigated agriculture system, and Laxmipur hill block. In Jeypore, buffaloes are required to help prepare the paddy fields, while in Laxmipur, Kondhs have a religious affinity towards the animal. Besides, it is able to escape from tiger and leopard attacks.

The primary aim of owning large holdings of cattle and buffalos is to secure manure. In Laxmipur, the sale of bullocks is a good source of income. Moreover, dead and old animals provide meat to the tribe.

In most cases among the hill tribes, during the dry season, the practice of herding cattle is not followed. After all chief crops have been harvested, cattle is left free to graze. While this practice is followed in Laxmipur, Rayagada and Malkangiri, in other blocks, people normally keep an eye on their cattle.

In other ways, the fodder spectrum primarily determines which forms of animal husbandry predominates in permanent cropping systems, and also the extent they are carried to.

Having examined in detail the relationship between the size of a family and their landholding to its livestock and its size in four blocks of Koraput, of which two are located on the hills and two in the plains, the following inference may be drawn:

Among tribals, poultry rearing is independent of family size and landholdings. Only in Koraput block does have a high correlation with the size of the family and the landholding. Some poor families rear chicken in their backyard, but no so the poorest of the poor and the labourers.

Goat rearing showed high correlation to size of family and their landholdings, which indicates the risk bearing capacity of such families to goat losses if appears.

The livestock economy matters because of its ready availability throughout the year. It ensures some income to sustain a family during times when their lands have to be left fallow.

Family size had high co-relation to number of cattle, bullock and buffaloes in home, whereas it was not so for number of sheep, goat and poultry. In Kundra block, however the number of sheep in families showed high correlation to family size, but not for landholdings. Thus it is evident that the tribes in Kundra are increasing their share of income from sheep rearing independent of agriculture. This is the scenario under land limited rainfed agriculture conditions, where farmers look for other sources of income particularly animal husbandry.

Bio-diversity in Livestock, Poultry and Disease Resistance

The livestock and poultry population in major, isolated groups of tribes show physical characteristic variations that are indirectly related to ecology variation, so also the production characteristic. From their perspective it is useful and productive in many ways but may not be so when compared with improved varieties. It is undisputed that no other improved breed except in few cases, can survive and breed effectively with the present feeding and management conditions prevailing in these societies. After 14 to 15 years of service in all the spheres of modern dairy management systems in Central Cattle Breeding Farm, Sunabeda, I have found that local hill tribals are not inclined towards cross-breeding the local cows with Jersey bulls because of survival problems and the cumbersome management process. Small ruminants and local poultry population meet the immense requirement of the tribal life, which has been described in the respective chapters.

The tribal livestock adapt quickly and show high resistance to many diseases, ecto and endo parasites, they live in the open, under the hot sun and during severe winter, and even pass nights while standing on wet mud. These animals are never treated with modern medicines. They thrive and reproduce under scanty grazing conditions, survive during scarcity, and show a remarkable rhythmic breeding activity to the availability of fodder under natural conditions. They are never home fed or supplemented with mineral nutrients, vitamins or any antibiotics.

The tribals show wisdom in stratification of livestock. The grazing system, special techniques in breeding and animal production have been described in various sections.

Cattle and Buffalo in Tribal Farming System

The cattle outnumber the buffaloes. Owing to religious links in their culture, the Kondh, Gadaba, Dongria Kondhs and some sections of Poraja tribes show special affinity towards buffaloes.

They have the following utility in tribal life:

- Most tribes like Bondo, eastern Koya, Gutob Gadaba, Dongria Kondh and Porajas do not milk cows. They leave the milk to calves.
- Manure from cow-dung is of most interest.
- They do not use cow-dung for fuel.
- Both cows and bullock serve draft purposes in agriculture operation; it is common for most tribes.
- The buffalo bullocks are preferred as the animal of choice to plough wet land; this is due to the buffalo's anatomy and its fetlock joint and hooves.
- The old and sick animals are used as food value among beef eating tribes.
- The buffaloes serve the Gutob Gadaba and Kondh for their Gotar and Meriah ceremonies, respectively. The Dongria Kondh sacrifice buffaloes in most ceremonies. They are vital for the religio-magic ceremonies they need to follow.
- The cattle is preferred over buffaloes because they breed more prolifically.

Koraput district has the Manda buffalo breed, found around Kakrigumma. Paralakhemundi buffaloes breed in Gunupur area and the Kashipur Mahisi (Kalahandi breed) in the Kashipur area. The Motu cattle of Malkangiri and a milch cow strain of Gunupur with various small herds of inbred cattle, are reported. They were described by Kornel (1999b).

Plate 131: Goudo women selling milk product, Boipariguda

Plate 132: Lampataput Goudo breed of cattle

Plate 133 Buffalo bullocks puddling paddy field, Kotpad

Plate 134: Cattle threshing paddy, Ksakrigumma

Plate 135: Cattle herd with heap of cattle dung, Ganjam

Plate 136: Cattle bells, Ramgiri

The cattle is non descript, with a short, prominent hump. This hump is less prominent on the animals in the plains.

Feeding and Grazing

They live solely on grazing. They can withstand the late summer heat and even scarcity of grazing materials. They put on weight during the rains and lose it in the summer.

Some of these animals have a common salt deficiency problem, particularly the bullocks purchased from coastal belt. The farmers in the plains take care of their bullocks, and feed them common salt.

Breeding and Reproduction

The bull calves are castrated when they are about two years old, so as to encourage good muscular development of the neck and the hump. These bull calves serve the herd. There is no selection of breeding bulls. However, the buffalo bull calves are castrated early to avoid viciousness.

Many still follow indigenous methods of castration. The cows usually exhibit the seasonal trend of breeding. The breeding season starts before the monsoon. The cows do not show regular annual calving and the maturity period is very long.

Diseases and Health Problems

It is common to find ectoparasites on the bodies of the cattle and buffaloes. Flies during monsoon and ticks during summer are also common. Endoparasites are also a usual menace. The Kondhs and Goudos use herbs to treat roundworms in buffalo calves.

Up to the 1960s Koraput district, cattle were used as pack animals to transport salt by *baiparis*. Somehow, the practice survived in Gunupur region.

The acidic pH soil also contributes to the poor health and is manifested as deficiency of phosphorous, zinc and calcium.

Cattle and buffalo deaths usually take place during mid-monsoon. In some areas, the foot and mouth disease or anthrax or trypanosomiasis are reported. Hemorrhagic septicemia and black quarter are common throughout.

During heavy rains, they spend the entire night standing on wet mud. The cold wind and poor grazing conditions are also responsible for their death, especially on the hills.

A large number of bullocks (cattle and buffaloes) are sold to Koraput from markets in neighbouring Andhra Pradesh. Middlemen, belonging to a scheduled caste, play an important role in this business.

Simultaneously large number of old, sick and culled animals are sold in weekly markets of Nabarangpur, Jeypore and others, from September to May. Many animals also come from Kalahandi district to these markets.

The tribals living near towns and on main roads located near villages know how to milk cows and have even taken to selling milk.

At present, the off take from cattle herds is below 5 per cent, whereas in goats it is 25 to 40 per cent.

Sheep and Goat in Tribal Farming System

The tribes have a special affinity for small ruminants in particular goats and sheep, which can also be seen from the folk stories (Kornel, 1999b). Sheep and goats are very intimately associated with tribal life through religious ceremonies. The colour of their coat has much to do with this. Similarly, there are certain characteristics, such as beards in goats and wattles in both sheep and goats, which disqualify them from ceremonial purposes. Should a goat climb the roof, it is feared that misfortune will befall the family (Kornel, 1999a). They have many reasons to include these small ruminants as important components in their age old, traditional, mixed farming system.

Oberoi et al. (1992) found that goat rearing was more remunerative than sheep rearing in western Himalayan tribal villages. Salem and Hammonda (1995) studied goat raising under traditional system in Tunisia. They found the results of goat production were better on small farms rather than on larger farms.

Goat farming in Koraput district is normally a secondary interest and as such a backyard operation. Traditionally, goats have been kept by tribal farmers as a ready resource for cash, to offer in ceremonial functions and for chevon on special festive occasions.

- These small ruminants convert otherwise valueless resources such as crop residues, forage from marginal lands and forests to animal products.
- The manure produced is utilized by crops.
- They also serve as a cushion in the years when crops fail. They are sold for cash when an emergency or special needs arise. The young people meet their expenses for their recreations, such as cock fight, dance or mythological plays called 'Nata'.
- They are made to walk in flocks over small, freshly sown millet fields, to give suitable depth for the tiny seeds to germinate.

- They are offered as sacrifice, at times, for various religious ceremonies, festivals and village feasts.
- They stand as symbols of status and wealth for a family and are presented as gifts between families of bride and bridegroom during marriage ceremonies.

Production Traits and System of Goats and Sheep in Tribal Villages of Koraput

- 5 to 14 per cent of the households keep goats and sheep.
- There are between 1 to 20 goats and / or sheep per household.
- Goats are more common on hill forests, whereas sheep are common in the plains.
- Goats are raised for chevon and sheep for meat production.
- Goat and sheep graze along with cattle and buffaloes in herd(s).
- The village community employees a grazier (family) on an annual payment basis, with daily supply of cooked food. In Laxmipur and in hill blocks, owners take small flocks for grazing.
- The new born kids and lambs are taken care of by women; men bring tree lopping to feed the mother doe and ewe.
- The women report the loss of goats and sheep, if any, after they return from grazing to men.
- Women participate equally with men on decisions on sale of goats and sheep.
- No concentrate, salt or mineral mixture or stall feeding is provided to goats and sheep at home, except for their young ones.
- That goats breed prolifically proves profitable.
- The sheep wool harvesting and processing is unknown.
- Surplus uncastrated males that are one year old, fetch the best price and are sold.
- Old does and ewes are generally sold. Sale of surplus sheep and goat brings profit.
- No selections are made to improve performance of the flock.
- The goat and sheep are not dewormed.
- Predators are common to sheep and goats in villages.
- Goats are believed to be less disease resistant than sheep.
- That goats tend to runaway, while sheep stick the flock, makes the tribals favour the latter.

Plate 137: Koya ceremonial buck

Plate 138: Goats with wooden neck bells

Plate 139: Sheep and goat flock walking over newly sown small millet field

Plate 140: Sheep housing in Kundra

- Grazing is the only expenditure for the family.
- Mortality, theft and predators account for loss.

The goat biodiversity of Koraput has been described in detail by Kornel (1999b).

The Goats in Tribal Home

There is considerable debate over goats as potential destroyers of the forests. However, research is needed to establish the truth. The tribals are highly conscious in regard to management of forests around their habitats, and the ecology around their life. It has been observed that the goat population amongst tribals living on the periphery of the forest are small and of dwarf type, which graze lower level weeds and grasses or bushes. The Barbari breed of Uttar Pradesh, Black Bengal goat breed of Orissa also fall in this category. These goat adapt well to stall feeding also. In comparison to the plain breeds like the Ganjam goat breed and others which are leggy as well as aggressive graziers due to their ability to stand on their hind legs to browse bushy plants, these are less harmful to the growth of the forest.

It is assumed that the population of small ruminants particularly goats will not grow a lot because the tribals, for the reasons explained earlier will not specialize in the near future. Moreover, as they keep goats inside their living huts and there is not enough space left to accommodate more. The tribal villages inside or near the forests are very fond of goat rearing but losses due to predators like tiger, leopard, wild dogs, fox and hyena have discouraged the system.

There is a new trend where tribals particularly of plain areas, are investing more in agriculture and horticulture. This system is gradually reducing the number of goats in villages and increasing the number of sheep. They also acknowledge the feuds and quarrels appearing on account of the errant goats and their grazing behaviour.

MANAGEMENT

The goats are grazed in community grazing lands along with cattle, buffaloes and sheep. They graze during day time and are penned during the night, to save them from predators and theft. The farmers do not supplement poor grazing, even in the dry season. Small farmers of Laxmipur and Rayagada area, however, follow the tethering system of management, whereby 3 to 10 goats are grazed in a group along the shift cultivated land patches on hills. Children and cheap family labour is made use of to manage the goat.

Farmers in Laxmipur, Koraput and Nandapur blocks pen the goats in the entrance room during night. In Kundra, Boipariguda, Mathili and Malkangiri regions, most farmers house goats and sheep in separate pens, with raised floor. It is made up of wood planks and bamboo. This is a common practice in Bastar district also.

After the harvest of *kharif* crops, goats are also left free for grazing without graziers in Malkangiri, Mathili, Korkonda, Laxmipur, Narayanapatna hills, Kashipur and Rayagada. They graze throughout day in small family flocks, and return back home in the evening. The Maria tribes of Bastar also have this system, even to the extent that they do not pen the goats. It has been felt that this system of goat husbandry is injurious to development activities pertaining to plantations. The goat breeds in the district are Malkangiri, Raighar, Narayanapatna and Koraput hill goats. They weigh between 20 to 35 kg. Kornel (1999b) has reported in detail the goat bio-diversity in Koraput district. All goats are prolific. They mature at 6 months of age. First one lamb is born, followed by twins. Triplets are not uncommon. The poll gene is absent. The Malkangiri and Rayagada farmers do castration of male kids but not followed by all in the district. The goat mange in the 1980s was a severe setback to farmers, followed by PPR. The contagious ecthyma is a major disease of young ruminants.

The Sheep in Tribal Home

Chouhan et al. (1991) showed that among sheep farmers of Himachal, poverty was severe for small farmers. It was attributed to low income. The sheep population here belongs to the thin and short tailed group as found in peninsular India. Ewes are medium sized and their body weight varies from 16 to 30 kg. The legs are normal in length and not leggy. They are long necked and the ears are medium, but sometimes they are crunched too. The ewes are polled, rams are horned or polled. The hairy ones resemble goats.

The body colour varies from light to reddish-brown (tan) and, very seldom, black. The face, legs and the belly of the sheep remain free of wool. Fleece, if any, is of the same colour as the body. There are hairy as well as wooly sheep in the population.

The fleece may be 6 inches in length and is moulted. In this area people are unaware of wool and, therefore, unaware of shearing or its varied uses.

ADAPTABILITY, BEHAVIOUR AND TEMPERAMENT

They are hardy and well adapted to a high monsoon rainfall (June-October) tropical environment, in which they have been reared for centuries. They adapt very well to direct sunlight of summer and exhibit 'head below abdomen of mate' phenomena to protect itself. They adopt very well to heavy rainfall conditions.

Sheep show good maternal behaviour and are normally excellent mothers. Their milk production is just sufficient for single lamb.

GROWTH

There are no recorded studies in this aspect. It is apparent that lambs grow well and gain body might up to the age of 9 months. The tendency of some lambs to grow faster indicates that there is tremendous scope to their improve growth rate with proper nutrition, management and selection. The body weight of adult ewes varies from 18 to 30 kg.

HEALTH

The health status of sheep depends on the intensity of numbers, housing, rainfall and humidity. Farmers claim to prefer sheep over goats because of the former's superior health.

Examination of the slaughter house indicates the existence of liver flukes, ruminal flukes and also other worms like haemonchus. The extent of the latter's involvement is not studied yet. The fact that sheep are not dewormed here, is proof of their remarkable resistance to internal parasites.

Other diseases appearing are mange, foot rot, pneumonia and gastro-enteritis, which may have parasitic origins. Myasis is not a problem, as encountered elsewhere among exotic sheep in India.

The enterotoxemia as a cause of lamb mortality has not been witnessed here yet. There is no incidence of sheep pox or blue tongue either. Blue tongue is highly prevalent in the plains. The most concern is over the emerging Peste-de-petitis (PPR) epidemics, resulting in high mortality among small ruminants. There is more to be done in the field of disease identification in sheep of the area, and for adoption of better preventive and treatment measures.

LAMB MORTALITY

The farmers do not complain about lamb mortality as a problem in their discussions. With limited supply of milk, multiple births do cause lamb mortality. Perhaps because they consider this to be natural, farmers do not pay much attention to lamb mortality. Studies are needed in this aspect.

ASPECTS OF FEMALE REPRODUCTION

The sheep population here show high variation between flocks for most economic and reproduction traits. They mature between 7 months to 1 year and a first-time pregnancy always leads to the birth of a single lamb. Most sheep produce only one lamb but less than 3 per cent may have twins. Birth of triplets is even rarer. In a period of two years, sheep can produce three lambs. This suggests that they breed quite prolifically. They have genes for multiple ovulations and with proper selection the population of sheep can attain more prolificacy.

Another important feature is the ability of sheep to breed throughout the year, or, at least, to have more than one breeding season. Lambs are usually born in November-December, and sometimes during February-March. Lambs born in the latter period, however, have to deal with the problem of inadequate milk from their mothers. It has been observed that, within a month of giving birth, ewes are ready and return to fertile breeding. With good nutrition and proper management, sheep can regularly have three pregnancies in two years.

MANAGEMENT

Sheep farming in Koraput district is normally a secondary interest and as such a backyard operation. Traditionally, tribal farmers have kept sheep as a ready resource for cash, to offer in ceremonial functions or to use as meat for special occasions. Of the total sheep population in a village about 85 per cent come from farmers who keep two to six sheep.

Farmers usually do not deworm their sheep. They do not realize that regular deworming can make the sheep highly productive. Lambs are allowed to run with their mothers and suckle as long as the ewe will allow. The newly born lambs stay under the care of women at home, where the ewe is provided fodder. At noon, women take the lambs to be fed on their mother's milk. Young lambs are never castrated.

In order to make the livestock rearing more profitable, they need to be more productive and efficient in feed conversion.

Tribal Pig Rearing at Home

Pig rearing is tribe and caste specific. In the tribal society of Koraput, particularly in the hill blocks, almost all households raise pigs. The tribes like Kondh, Sano Poraja, Bodo Poraja, Koya, Gadaba, Dongria Kondh, Bondo, Pengo, Ollar Gadaba, Didayi, Parenga and the scheduled caste communities raise pigs at home.

The pig rearing is in diminishing trend due to

- Growing interest in agriculture.
- Increase in multi-ethnic groups in villages.
- Medical advice during suspected Japanese encephalitis epidemic and human deaths.
- The fear of feudes due to damage of paddy fields and other crops.
- General understanding that the farmer whose crop has been damaged by a pig, has the right to kill it on the spot. This is not considered to be an offence.
- The carcass of the dead pig can be retrieved only after a payment of Rs. 500 to the owner of the damaged crop.

The pig husbandry at tribal homes has the following features:

- They mature early and grow fast.
- One sow for each household.
- The sow is for production of piglets, may be 6 to 12 per farrow.
- Under good management, sows usually produce two litters per year.
- The piglets get some milk from the sow in most cases, which helps piglets to survive the first 2 to 3 months.
- The family gradually removes piglets from the litter. This is done to enable the sow to rear the remaining piglets well with the available milk. The piglets that are taken away from the litter are either sold or consumed.
- At the milking stage, the sow becomes very weak and lean and her ribs become prominent.

- Farmers do not rear pigs to become adults, unless they need replacements for the sows. They are well aware of the limited feeding options available to adult pigs.
- They are very well aware of the capacity and the number of pigs they can sustain under scavenging conditions in the village.
- Pig styes are close to their own living quarters, with strong wooden gates to protect the pigs from predators during night.
- The pigs are reared under free-range system in the village itself. But during the *kharif* crop season, i.e. from 15 June to 15 September, farmers keep pig indoors, with a fence around their stye, and feed them little rice bran to live. This is evident in Kundra block, and Sano Poraja villages. Most villages in Koraput block sell all pigs during the cropping season and buy them back to rear from October onwards.
- Feeding troughs are made out of wood or stone to feed the pigs.
- Pigs are fed rice bran and the water used to wash utensils, to feed on.
- The carcass return is quite high and is between 65 to 80 per cent of the live weight.
- They acknowledge the profit from pigs.
- The women remain in total charge of the pigs at home, and keep an eye over them.
- Predator losses are encountered.

Pig Breeds in Koraput District

There are two distinct types of pig breeds found with the tribals and scheduled caste populations. The local pigs are called *Burudi* in Koraput. They are all of the *pot bellied* type. This breed of pig is distributed all over in the local tribal villages. They weigh about 30 kg. They are generally light black, although some have white patches on their bellies or legs. The tail is long with tufts of hair, hanging at the hock. The face is long, with a full jaw. They scavenge around the village, and generally do not go out. They have little back fat, which starts accumulating after they are a year old.

They reach early maturity at about 6 to 8 months. They farrow twice in a year. They can produce up to 12 piglets per litter but, usually, the number is between 6 to 8 piglets. A large, well fed sow will produce

Plate 141: Transport of village pig

Plate 142: Poraja women feeding pigs at home

the highest number of piglets. Males mature earlier than females, and the sight of an adult boar in the tribal villages is uncommon.

Young males are, therefore, used as breeders. There is no selection of any kind and at many places, inbreeding is seen. The early winter period, October to November, is the most common farrowing season. Some litters have brown lines spread over their backs. This is a sign of their link with their wild relatives.

These breeds show high fertility and seem to be more independent of feeding conditions.

Another breed mostly seen in the urban areas of the district, particularly in the Rayagada belt, is called *Jhinga* or *Pondi*. They grow up to 50 kg. They have black skin and a barrel shaped abdomen. The tail is thick at the base, and twisted with small tufts of hairs. Some boars have long, standing hair at the neck.

They are equally prolific, and are preferred in places where feeding conditions are better even under a scavenging system.

They have a good herding ability, and move in batches under the supervision of one grazier or caretaker.

However, under low-scavenging or semi-scavenging conditions, the local pigs are superior in terms of survivability and, thus, production. The feed situation in the villages will not change dramatically because it is related to changing activities of life, which is spent in pursuit of food production.

The tribals well understand the limitations in pig rearing and have, therefore, evolved their own systems, which are based on their local breeds.

In almost all villages there are local surgeons who perform castration operations on piglets by open method.

Pig Mortality

Piglet mortality is about 30 per cent in village conditions. This is mainly due to pneumonia caused by lying on the damp floor of the shed, and inclement weather, diarrhoea, piglet anaemia and, sometimes, by being crushed under their mothers.

Swine fever, a viral disease described like the Ranikhet disease of poultry (Kukuda marudi), causes serious loss. Foot and mouth disease (FMD) is also seen among pigs when outbreak occurs in cattle; hyper

Plate 143: Koya tribe pig style

Plate 144: Poraja tribe collecting waste food in a pot for pigs

keratization with skin irritating is seen among hill village pigs of Koraput. This may be also due to zinc and vitamin A deficiency.

Pig mortality and poor growth due to round worms is a major problem, as is the loss to predators in almost all villages.

Village Poultry in the Tribal Farming System

The village poultry production systems in India and also among the Koraput villages, are based on indigenous or native domestic fowls (*Gallus domesticus*), wherein the hen brood and hatch eggs and rear chicks. This existed in the villages from time immemorial. All tribal groups and Scheduled Castes keep poultry. Kitalyi (1998) has shown that 70 per cent of the rural poultry population is of the indigenous type, kept on low-input and -output production system in Africa. It is 94 and 90 per cent of the total poultry production in Nigeria and Bangladesh, respectively (Sonaiya et al., 1998).

They form a common resource for all the different gender groups in tribal villages, to which the landless and also to those who do not possess cattle, sheep, goat and pigs, attach high socio-economic value. The village poultry system should be placed high in the food security programme. They are often the only source of animal protein for resource poor households.

In tribal village of Koraput district, 53 to 76 per cent of the households are found to keep poultry at home. The number of poultry per household is significantly different between the villages on the hills and in the plains. Moreover, the households who solely depended on wage earnings have more or less no poultry at home. The same is the case with the families of the community livestock graziers because there is no responsible person at home during the day to take care of chicks or adult poultry.

The distribution of poultry among households in four blocks of Koraput (1995) are shown in Table 10.

Tribal Village Poultry Production System

The three poultry management systems, intensive, semi-intensive and scavenging are differentiated on the basis of flock sizes and input-output relationships (Sonaiya, 1990 and Kitalyi, 1998). The extensive

TABLE 10: DISTRIBUTION OF POULTRY AMONG HOUSEHOLDS IN FOUR BLOCKS OF KORAPUT

Household	Blocks			
	Jeypore	Kundra	Koraput	Laxmipur
Percentage keeping poultry	53	63	62	76
Average poultry	9	7	4	4

production systems in Africa, where the poultry is kept in free range or scavenging, is different from the more recent extensive free range poultry coming up in developed countries (Thear, 1997).

The tribal village poultry production depends on scavenging within the household boundaries, and then within the village boundary. Although the village poultry move freely about the whole village, they are all attached to a specific household. It is not imperative to say that village poultry as part of the household often share the same shelter.

A study by Ramlah and Kasim (1992) showed that the rural farmers in Malaysia mainly practised the free range system (82.1 per cent).

OWNERSHIP, WOMEN AND POULTRY

The village poultry is under the charge of women. While the general use and sale of cocks depends on men, women definitely have a say in the matter. The women go to the weekly market for sale or purchase of poultry. Gueye (1998) reported that 70 per cent of the poultry owners in rural areas of sub-Saharan Africa are women.

The ownership pattern of village chicken production system in Africa was reported by Kuit et al. (1996), Williams (1990) and Kitalyi (1996). In Ethopia, Kenya, Uganda and Tanzania, three pattern were discussed. They are family ownership, by men and women and individual ownership. This depended on the mode of acquisition-purchase, gifted or inherited. The ownership of poultry in African societies is a product of social and cultural aspects of society.

The village poultry husbandry and production system with relation to household and inter-household in villages and between tribes do differ. In the tribal villages of Koraput, poultry flocks of all households scavenge together but are fed separately, by individual households.

The Bhattra, Omnatya and Bhumia tribes prefer to have a large backyard with clay boundary walls. The poultry production system here is single family flock. But among Sano Poraja, Bodo Poraja tribes and others it is mixed family flock system and poultry scavenge all around households. Among the Kondhs, the poultry has to remain behind the house, as described earlier. They have parallel close running rows of houses with a open front street. The backyard has cattle byre, a pigsty and large open yard of manure pits. People like to keep the streets in front of their houses clean and painted, and free of pigs and poultry. Therefore, the backyard manure pits have to serve as the feeding grounds for the poultry and pigs.

The management of·poultry in the village is complicated by the presence of multi-age groups in the same flock. This is unavoidable in a scavenging system.

SEASONAL TREND AND POULTRY REARING

The village poultry production shows seasonal trend among these communities. It starts with hatching of eggs from September end and continues up to March end. The hatching operation is at its maximum during winter months. In the first and second weeks of April the tribes celebrate the annual Chitra Parab, during which period a fair amount of poultry is roasted at home. As a result, except for one or two hens, the entire flock is used up. This serves as a kind of preparation for the coming agricultural season, saving any damage to their backyard farms, which constitute their food security. To stop the poultry from entering these backyard farms, sometimes, their legs have to be tied.

POULTRY PRODUCTION SYSTEM AND CHARACTERISTICS

The village poultry production is characterized by low level of inputs and outputs. The main feed comes from household refuse and scavenging. The other inputs are housing, additional feed from rice by-products, which is depends on the economic circumstances of the household.

With such low inputs, or near zero contribution of the household except for small labour contributed by women at home the village poultry is capable of generating cheap and readily harvestable meat and eggs for immediate home consumption and income generation.

Plate 145: Dongria Kondh youth with ceremonial poultry

Plate 146: Cock fight

Plates 147 and 148: Broody hens sitting over hatching eggs

POULTRY PRODUCTION TRAITS

On an average the tribal families maintain 2 hen units per family. Hens, on an average, lay 10 to 15 eggs per clutch. However, 20 eggs per clutch is also frequently seen in Kundra, Boipariguda and Mathili regions. The hatching percentage is above 80 and even up to 100 per cent. The chicks mature at 5 to 6 months of age.

NEW HATCH MANAGEMENT

The chicks are given clean water after hatching, and fine grounded rice. A circular, portable pen made from bamboo and called 'thapa' is placed over mother hen and the chicks are fed. It is also to prevent predator attacks. The local poultry havc strong mothering instincts.

FEED

Rice bran and ragi flour are the main constituent of the limited feed supplied by family. Depending on the economic status of the family, feed is offered once or twice a day to the poultry at home. Their protein requirement is usually met by feeding on insects.

POULTRY TYPE AND BREED

The local poultry population in the district in different locations shows clear and distinct characteristics. Selection pressure continues in some of the blocks by the villagers. This, therefore, shows variation in terms of shape and size and some phenotypic characteristics like comb shape, size, plumage colour, also height and body weight. The bio-diversity in poultry population, like fizzled feathers, back lock wings, different plumage colour, necked necks, comb size and shapes, etc., is maintained by linking the need to various festivals and ceremonies of life (Kornel 1999a). There are three breeds of poultry in Koraput district—Vezaguda breed, Dhinki breed and Kalahandi breed.

DISEASES AND MORTALITY

The most limiting factor for the village poultry production is the Newcastle or Ranikhet Disease Virus (RDV), which often accounts for 80 to 90 per cent mortality. Another major disease is Fowl pox, which affects chicks. This causes blister like lesions on skin, specially on the face of little chicks. The chicks often lose their eyes and most of them die.

Most vaccines available for Fowl pox are to be administered when the chicks are between 6 to 8 weeks old. But, in an infected village, most chicks are already dead by that age.

Earlier a sort of seasonal incidence of RDV was reported, but now it is happening throughout the year. In certain places, roundworms and tape worms among chicks are also a major reason for mortality. Poultry lice is a major irritant for families, as it crawls all over the house and even on family members.

POULTRY AND UTILITY AT TRIBAL HOME

Up to 50 per cent of the poultry is meant for sale in the market, while the rest is consumed at home, usually on festive occasions. They fetch a good competitive price as demand for village poultry is higher than the commercial broilers. The roosters fetch very high price in winter, as demand for cock fighting is high then.

Tribal farmers keep poultry for hatching and the chicks thus produced are reared for meat. That, apart from cock fighting, is the main reason behind keeping poultry at home. Although they do eat eggs nowadays, specially when the poultry are under threat from a disease and it is considered risky for new born chicks, it is not very common for them to do so.

With the shortest gestation period among all the livestock the poultry help families overcome animals losses sustained during epidemics.

COCK FIGHTING

Fond of cock fighting, men hanker after roosters suitable for the sport. The roosters are tied with sharp knives at the spurs of their feet. Although it is illegal, cock fighting takes place almost everywhere on a weekly basis. Crowds start assembling from noon, while the bidding and fighting starts around 3 p.m. The rooster that looses is killed by the victorious team and is considered their property.

CAPONIZING COCKERELS

The Kondhs living close to the Kalahandi district practice ‘caponizing’ of cockerels, with their indigenous surgical technique (Kornel, 1999b). The same tribe, as also most other hill tribes, catch wild fowls, by various methods from the jungle for food.

As the value of poultry increases over time, it is concluded that traditional family poultry represents the investment, insurance and saving of a family. It also provides employment and disposable income to tribals, particularly in the off season. It is well integrated with other farming activities as it requires little time and investment.

Tribals and Duck

The tribal villages of Koraput, particularly around the Pangam river of Mathili, the Kotpad and Nabarangpur area of Indravati river, have a good population of ducks. The ducks are called 'Kado hansa' or 'ducks of the mud'. The Bhumia, Bhattra, Matia, Dhulia tribe and Mali people of Koraput breed ducks. A systematic study of Koraput duck breeds was taken and reported by Kornel, 1999b.

In shape and size, the ducks are very similar to the Peking duck breed, specially white ones from Koraput. The ducks of neighbouring Kalahandi are smaller in size. The following distinct colour patterns amongst the Koraput ducks are seen: white, black and white, khaki type and bluish black type.

All these ducks are well known for their capacity to lay eggs. A study is needed to report the production aspect of these duck breeds. Some claim that the ducks lay more eggs than local poultry, seasonal in egg laying and some claim that ducks lay eggs on every alternate day. The tribals eat duck eggs as well as duck meat; ducks are also used for ceremonial purposes. The Kalahandi Kondh tribe and others breed the female duck with the male of Muscovy type. The hybrid thus produced is called 'Bodhia', which is known to be sterile. They grow fast, with large quantity of flesh. The duck eggs are hatched by hens.

Muscovy Ducks

They are called 'Moti hansa' and are well distributed in the plains of Koraput including Malkangiri and Nowrangpur. From small to large in size, their body weight is between 4.5 to 6 kg. A family flock may be between 2 to 20 Muscovy ducks. They have black, white, black and white mixed, and dull brownish-black plumage patterns. They are bred for their meat only. They are home fed on rice bran and the water used to wash utensils by families. They hatch their own eggs, at least twice a year and there are between 8 to 12 ducklings in a clutch. They are least harmful to backyard farming.

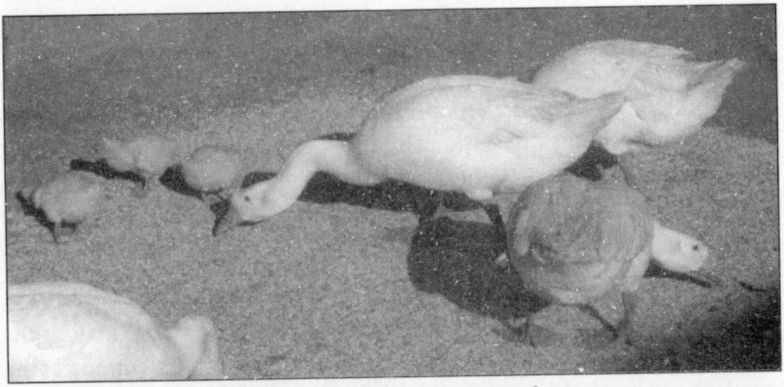

Plate 149: Rajhansa, Kotpad

Plate 150: Geese, Papadahandi

Plate 151: Muscovy Duck, Malkangiri

Rajhans (Geese)

They are specially found in Bhattra, Bhumia and Halwa tribe villages in the plains, particularly around Indravati river of Nabarangpur and Kotpad region. They are large in size and have a white plumage, and/or brownish or khaki patches over their wings. They are considered to bring peace to the village. They hatch their own eggs.

Pigeons in Kondh Home

The Kondhs and Koyas keep pigeons at home. They make suitable arrangements to keep them safe from predators. The pigeons are sacrificed in special ceremonies and functions of the Kondh, Dongria Kondh and Koya tribes.

Wild Animals as Pets in Tribal Home

The wild piglets find a place as a pet animals in tribal homes. They take care of them by feeding them at home. But when they grow, they become vicious and are killed for pork. There is no taboo attached with eating wild pigs, even for those belonging to high castes.

Wild poultry particularly adult roosters can also be sometimes seen in Kondh household. They are kept tied inside the house during daytime. They are used to catch wild fowls from the forest. Similarly, wild partridges, known as 'gundri', are caught from harvested fields snared in special nets. Some of these are reared at home. Gundri meat is delicious.

Feeding and Grazing of Livestock in Tribal Villages

At the onset of monsoon rains, the dry plants, particularly the doob grass, are immediately revived to green. The continuous rains thereafter support green vegetation until November. During this time, plant growth is optimum. December onwards, the rains stop and the grass and vegetation become dry and grey. The plant growth stops. The feeding resources for the livestock gradually diminish. It gets worse in April-May, when there is a scarcity of feeding material for the animals.

Paddy Straw as Feed

The Malkangiri farmers, harvest either the crown of the rice plant and leave straw in the field or, after threshing the paddy, they leave the straw on the floor. The cattle herds feed on these heaps of paddy straw.

As usual, paddy straw contributes some cattle feed during fodder deficient months of the year to plain land farmers, and few rice farmers of the hills. The hill farmers mostly depend on the ragi straw and other bio-products of crops to feed their livestock. Conventional grazing and a meager amount of paddy straw, during the lean summer months, keeps the animals alive.

Conservation of Paddy Straw

The hill farmers have the age old techniques and feeding schedules of paddy straw to their livestock. In some tribal villages, they stack the paddy straw in a radius of 18-20 ft and between 20-25 ft high, and cover the crown carefully so that no rain water percolates. A stone circular wall, all around the stake saves it from the entry of stray cattle. This is common in Koraput, Nandapur, Lamtaput, Pottangi blocks. Paddy straws heap up over four pillars, which stand about 4 ft height and represent another staking method particularly in Rayagada. Small stakes can be seen all around. They are very careful to feed paddy straw to their valuable bullocks, particularly during late summer months.

Ragi Straw

The ragi is harvested manually, and the straws are left back in the field during winter. The village herds are generally seen eating this straw. Compared to the paddy straw, the cattle find ragi straw less palatable.

The tribal farmers are not aware of the hay and silage method of conservation of fodder.

Feeding of Bullocks in Tribal Home

The male members favour bullocks. They feed them with paddy straw, green grass and rice-bran at home. They know the need of feeding salt to bullocks, but refrain due to the costs involved.

Feeding Sheep and Goats

Goats and sheep literally live by grazing alone. After they have given birth, however, mothers are fed branches of trees and bushes, like dumbri, etc., for 2-3 days.

Feeding Pigs

Families acknowledge that pigs need to be fed at least three times a day. However only rice bran is all they are offered in most households, along with rice decanted gruel.

Species of Livestock and Grazing Behaviour

The villages in the plains follow an organized grazing system, through village herd(s). This is usually a mixture of 80 animals, or less, in a single grazing system.

Cattle graze tender grass leaves, buffaloes go for the fibrous stocks, while the sheep graze the last leftover grass quite close to the ground. The goats prefer browsing of branches.

SEASON AND TIME OF GRAZING VILLAGE LIVESTOCK HERDS

During winter, grazing starts around 9.30 a.m. and the animals return by 4.30 p.m. In summer, the animals are sent out by 8 a.m. and return by 5 p.m. To avoid the hot afternoon sun, the animals are kept in shade, under trees, between 12 and 2.30 p.m. Grazing becomes very difficult during the long, rainy spells, when the hills are slippery and the streams around villages limit the herd movement to field.

The Kondh, the Koya, and many other hill tribals, except in Koraput, Nandapore, Jeypore and Nabarangpur area leave their livestock free to graze from to June-July. The goats and sheep roam in small flocks, and return home after grazing in the evening. This behaviour is unique. This has a negative effect on new plantation.

VILLAGE GRAZIERS

Village grazier usually belong to the Dom or Scheduled Caste community of the village, and may even be from the same tribe, This is a family occupation and members distribute their time for grazing responsibilities. It is a commitment they make virtually to the entire village. It is common to find elderly people, who are accompanied by one or two young children, looking after the grazing. In the Koraput block, by rotation, each family sends one member to participate in village herd grazing. In the Kondh villages, clans join together to form small units for grazing and participate on a rotational basis.

Sometimes on loss of animals from the herd due to predators or otherwise, the village graziers bear the assaults from the owners. They are the weakest economical section of people in the village.

SELECTION OF VILLAGE GRAZIERS

The village head and village members appoint the village grazier for one year.

PAYMENTS AND TERMS

The village livestock graziers usually belong to the very poor families of the village. By rotation, every household in the village offers cooked food to the grazier's family twice a day. During the annual festival in February, the families also receive annual payments in kind, in the form of paddy and ragi, etc., and some amount in cash as well.

In turn, the grazier offers some mutton or chevon and gift to each household. In appreciation of this, each recipient family, has to reciprocate in a big way. The grazier celebrates the event with dance and local music. The rest of the village also joins in.

OTHER JOBS OF VILLAGE GRAZIER

They report about incidence of diseases in the herd and sick animals to the owner families.

If there's a birth in the herd, the new born is taken care of and transported to the owner's residence carefully.

They are quite knowledgeable about plants and can distinguish between those that are edible, and can serve as fodder, and those that are toxic for the livestock.

The whole family of the village grazier is involved in, at the grazing activity. As a result, they do not even keep poultry at home.

If the animals under their care are lost to theft are or predators, or stray while grazing in the forest, graziers are blamed and have to face scoldings from owners.

Although they have a wealth of information and knowledge about breeding and the status of female animals in their herds, they do not find an appropriate place in any breeding programme.

With their knowledge and ability to handle the damage to the environment and ecology, soil fertility and the welfare of livestock, they are never consulted or made members of any such working group.

Grazing Land and other Resources

In every village there is some kind of grazing land called 'Gochar lands' owned by the respective communities. The gochar lands cannot be transferred by law for any other purpose than it is meant.

In some villages they graze in neighbouring forest lands also. The pressure on these lands from large stocking rate is quite evident through pronounced soil erosion.

During the rainy season, when all arable land is covered with crops and there is green grass every where, graziers are under tremendous pressure to protect the crops and prevent any damage or loss in the fields. It takes a lot of skill and constant observation throughout the entire grazing period to accomplish this. They usually graze animals by the sides of the roads.

In recent times, more and more unused land is being brought under cashew plantations and social forestry cover. As a result, fodder resources for the livestock are diminishing.

The only feed resources now are the 'doob grass' and some other type of coarse grasses. These are less nutritive and their total production per unit area is not very viable economically.

There is known technology gap both in order to improve the village community land resources and to produce economically more nutritive value fodder. At this stage of subsistence farming, the plantation fodder crop does not sound logical to tribal farmers. They believe that animals need to find their own feed.

LIVESTOCK GRAZING BY KONDH AND SABARAS

The Kondh tribe, clan and families take care of grazing of their own livestock on rotation basis where a representative of the family joins the grazing operation. They claim that grazing small herds on hills is very effective in order to gain good growth of animals, and efficient when crops are standing. The disadvantage in the system is that more labour days are spent.

NECK BELLS FOR CATTLE AND GOATS

During the grazing period, different kinds of wooden and bamboo bells are tied around the necks of goats and cattle. This helps them trace the animals, who wonder while grazing in the hills and in the thick forest.

Local Fodder Plant Resources

The local fodder plants, as denoted by tribal farmers are, Girli (*Indigofera pulchella*), Mamagua (*Vengueria pubescens*), Tendka (*Woodfordia fruticasa*), Ladan (*Heptaplelurum venulosum)*, Chelidantki (*Breynia patens*), Dumuri (*Ficus glomerata*), Chili (*Acacia concinna*), Mutrimal (*Smilax macrophylla*) and Khatakoli (*Zizyphus rugosa*). This needs further study. However, Thapa et al. (1996) have worked out the Nepalese farmer's knowledge base of tree fodder resources, and advocated research and development.

Plate 152: Village cattle, buffalo, sheep and goat grazing

Plate 153: Paddy straw for scarcity period feeding cattle conserved on raised pillars, Laxmipur

Plate 154: Paddy straw for scarcity period feeding cattle with fence around, Nandapur, Koraput area

Tribal Animal Husbandry System: Transition Trend

The traditional farming systems existing in various tribal societies, which are in different phases of development, are all subject to change and are in transition. In most of the cases though, livestock economy is important. It stagnates at a low level of productivity.

The livestock shall be the critical input for getting poor tribal people out of the poverty trap. This sector is showing growth and shall continue to do so in the development sector too.

Animal Production and Land

Due to population growth, migration of tribal population to different forests, mostly into reserve forest is the trend. It is for more pasture and agriculture land and forests. From populated hills, they are migrating to Ramagiri forests and Eastern Ghats, as well as to the Nabarangpur area. The movement away is being led by the Kondhs and the Scheduled Castes.

Already, the cashew plantation and coming up of coffee plantations in the district are attracting more and more immigrants who are affecting the landholdings and thereby diminishing land availability to the growing local population and for livestock is foreseen.

The additional livestock production, however, will have to come from already utilized land and land use intensification. In the latter, the provision of purchased inputs change the role of animals in most farming systems.

System Transition

- MIXED SPECIES TO SPECIALIZED SPECIES ANIMAL HUSBANDRY SYSTEM

The mixed species animal husbandry system of tribals is a self-sustaining system, whereunder constraints, particularly against the prevailing livestock diseases with no or negligible livestock service delivery system, the farmers feel secure to some extent.

The Koya and Bhattra tribes are in transition to species specialized animal farming. The Koyas prefer goat and cattle. The Bhattras prefer sheep and cattle. However, poultry is universal. The grazing of sheep and goats and cattle in separate herds is an indicator of this system.

SELECTION TO RETAIN OR IMPROVE THE EFFICIENCY OF INPUT CONVERSION

Efficient feed conversion to gain better profit ratio is the central theme on which the modern industrial livestock farming has developed. Further, continued improvement to gain more and more profit is the striving force in the industry. The research, extension and market have played important roles in development of industrial farming system, for example, the commercial poultry broiler and layers. It holds good in the case of all livestock species and the sector as a whole. The system is highly dependent on efficient genetic manipulation of animals.

However, such high efficiency is grain dependent. The poor tribal societies in transition shall not use such system *in situ*.

But the livestock species that have the genetic ability to convert roughages, specially poor type roughages, and those that can stand the stress of seasonal scarcity of food, and still gain faster growth during lush season, will need identification for further development through research. Saying the 'scavenging capacity' genetics shall be more sought in future can conclude this. The Koraput pigs as mentioned earlier, reproduce under poor scavenging conditions. The reproduction characteristic is thus independent of nutrition. The scavenging poultry breeds of the tribals will be the most sought after of species.

Species of Livestock and Future Trends

This shall depend on the following factors,

- *Disease resistance*: Losses due to susceptibility to disease, along with the mortality rate, will be the prime considerations before a farmer adopts a particular species of animal.

- *Productivity:* A prolific rate of breeding, which benefits the family shall continue to be a vital consideration. This will be the breed characteristics too.

- *Socio-religious factors*: Religious considerations, like coat or plumage colour, and type (like wattles among sheep and goat, hornless goat, beard in goat and long tail for the sheep, cattle without hump) shall be considered seriously by tribal farmers before adoption of any species or breed of livestock. The livestock species and traditional choice ceremonial offers shall continue in future without change.

- *Socio-economic factors*: The number of animals in the village shall be determined on carrying capacity and health of the environment, so both families with their interest, and community at large shall be the determinants. However, socio-religious system and socio-economic developments, in the future will see the removal of pigs from tribal families in the plains. The tribals will reluctantly do so.

  Besides the above four factors, the services available, market, access to credit and success stories from neighbours will affect the trend to species specialization.

TREND WITH CATTLE AND BUFFALOES

Traditional small holding systems often include large numbers of cattle and buffaloes to provide traction power, manure and fuel.

MANURE

In the course of transition and modernization, manure is lost to fuel energy. The tribals refrain from burning cow-dung for fuel. However in reality, there is significant increase in the number of cattle in Koraput district. This is to increase the total cow-dung manure for family farming, which helps to boost the dwindling crop production system. But larger number of cattle will damage the environment. And then, the chemical fertilizer will appear. This will reduce the dependence on manure to some degree.

This is a stage before the farmers enter into commercial base farming. Once that happens, there is a drastic reduction in the numbers of unproductive livestock. With increased land-use intensification, however, more farm by-products are available. Here, the chemical fertilizers widen the fodder based and allow farmers to maintain animals on very small land-holdings. Sooner or later, farmers will get into dairy farming, particularly in the irrigation farming system of Koraput.

Cattle, particularly the dairy variety, contribute to high yield of crops through manure, fuller employment and higher additional income. But this works best when there is an assured marketing system and ample provision of labour, which is hardly the case with rain-fed tribal agriculture. The scope is thus limited. However, limited scope for draft cattle with limited milk yield will be there on areas where irrigation facility is available. Generally, when farmers decide to go in for a milk market, the decision goes in favour of buffaloes. It is due to their superior adaptability to harsh conditions, efficient roughage utilization as feed and high milk fat preference of consumers. For paddy cultivation, particularly under the intensive irrigation conditions, buffalo bullocks are preferred. Somehow buffaloes have a strong socio-religious attachment with many tribal groups of Koraput, as mentioned earlier.

TREND TOWARDS SHEEP AND GOATS

The intermediate trend during livestock transition favours the husbandry of sheep and goats, with the latter coming up only when browsing lands are available. But when opportunity of browsing cease, it weighs in favour of sheep. The sheep replace goats when the farmers renew interest in agriculture and intensify horticulture. This was confirmed among tribals receiving water for irrigation from the Kolab dam in Koraput district. Due to their prolific rate of breeding, fast growth, price structure, readymade market and least input cost, the tribals prefer small ruminants in system transition. The issue of exploitation by middlemen needs to be addressed.

TRENDS IN VILLAGE PIGS

The authorities in development shall continue to place emphasis on cross-breeding of indigenous village pig population with large size exotic breeds to gain growth and body weight. But the poor tribals who are short in grain production shall not make any gain as the improved cross-bred pigs need good feed regime to be productive.

The present pig farming system in tribal villages, needs an effective disease control system, and shall improve confidence level of families to rear pigs. It should soon be followed by some amount of protein feed supplement to gain more profit from village pigs. The growth of pigs will give both cash income and cheap nutrition to tribal families.

TRENDS IN VILLAGE INDIGENOUS POULTRY

Single and most vital trait of brooding in indigenous poultry has kept going the self-sustaining poultry husbandry in tribal villages. Brooding trait of indigenous poultry get diluted or lost when they are bred with exotic improved poultry breeds and hence cannot hatch eggs. The scavenging behaviour, predator-escaping capabilities and high prolificacy of indigenous poultry will be most sought after in future poultry breeding. The tribals shall love to retain the qualities of poultry for game.

TREND IN BREEDS

The local breeds of livestock have come about due to years of natural selection, as well as some artificial selection. They are able to withstand a short but regular period of starvation, or scarcity of food, every year. The tribal farmers let their animals graze on their, to find their own food and survive. Once natural fodder is available, these breeds of animals gain weight and body fat. This, however, gets depleted during the time of scarcity and helps the animals survive. They also show a long period of growth curve compared to other, improved breeds.

These breeds of animals shall face upgradation of some kind with some superior indigenous as well as exotic breeds. The survivability of graded and crossbreds shall be the crucial fact for further adoption of the technique. This shall depend on the availability of animal health control measures, affordability of the farmers, and the availability of feed and fodder resources. Thus, economics shall be the major determining factor.

TREND IN ANIMAL FEEDING, PRODUCTION, LAND UTILIZATION AND ENVIRONMENT

However, in mixed species animal rearing systems, the free grazing in forests, and on communal and unutilized land continues until the environmental problems set in. However, the tribals intelligently balance the environment through animal species population reduction or increase, at the first instance. Lack of appropriate knowledge, adequate finance and further being unorganized they face constraint in the improvement of the degraded environment.

The key parameter for future animal production is apparently not land availability, but price relation of inputs like feed and fertilizer and animal products. In tribal markets, this is controlled by the intermediate traders.

If fertilizer use on grass remains uneconomical, as this is in most parts of the tropics, then animal production shall depend on, and remain restricted to the use of non-arable land and on by-products of arable cropping (Rutherberg, 1980). But if it is profitable, then practically unlimited feed resources can be made available at fairly constant costs per unit of output. This situation will assure profit ratio to input cost if genetic ability for grass conversion in the concerned animal species and breed is higher.

Under the general scenario, the trend goes from unimproved to improved pastures, unfenced or unprotected to protected farm areas; the establishment of artificial pastures; fodder conservation for the dry season; supplementary arable fodder production and in areas of great land scarcity, from grazing system to minimum of zero-grazing systems.

The Transition from Tribal Farming System to Industrial Farming System

- Probably two conflicting trends will dominate the change in tropical farming systems. Economic growth and the development of markets for inputs and outputs and the provision of technical innovations, which will promote specialization.
- Growing land scarcity, however, and the resulting trend towards land-use intensification, strengthens the interactions between crop and livestock enterprises. The impact of these conflicting trends on actual farming varies widely according to soils, climate and price relations.
- The gap from tribal farming system to industrial farming systems is long and the efforts to follow the latter by making changes in the former have always led to unsustainable development. This depends on many gradual changes, infrastructural developments and the need

of society. This will be the ultimate goal, which is 'process' dependent but not 'output' oriented.

The future emphasis shall include:

- Agro-ecology improvement; to utilize more animals in nutrient cycle (agro-forestry, rotation, manure management, etc.); gochar land management by villagers and high yielding fodder species for such marginal lands will require further research so as to improve the rate of adoption. This is a 'problem' area.
- The animal species of poor people (small ruminants, pigs and poultry) suitable for small holders. Genetic improvement of indigenous, disease resistant breeds; and controlling diseases like Peste-de-petitis (PPR), Blue tongue in small ruminates, African swine fever, and Ranikhet and Fowl pox in poultry.
- The other area of interest that will be vital for growth is to find an alternative to animal dung manure, thermo stable vaccines, low cost vector controls and appropriate and artificial insemination techniques, like room temperature semen, and cheap drug based heat synchronization of ruminants.

Ethno-Veterinary Medicine

There is a viable system of ethno-veterinary medicine, among tribals of Koraput district. The main sources of medicinal plants are from the neighbouring forests. Somehow, both pharmacology and medicine practice have stayed with one person, often becoming a family tradition. The medicine man is called Dissari. But it is clear from discussions on the subject in village forums that every individual, including women, have knowledge of the common medicines and specific uses. This is more so with hill tribes.

These medicinal plants are known among local tribes by specific names, with variations between tribes also. Medicine men are generally reluctant to disclose their knowledge, except for some commonly known facts. Perhaps this has something to do with tribal medical ethics.

The ethno medicine man is common for both veterinary and human medicine is one. The following few ethno-veterinary medicines in Laxmipur block villages, as practised by Desia-Kondhs are reported.

Against Intestinal Worms

The villagers in Talkaipadar use (1) Tuda fruit with milk (*Zanthoxylum aromatum*) to deworm buffaloes and calves. One or two fruits can be given to a calf. But a month old buffalo calf should receive a single dose only, followed by second dose at weaning. The villagers of Barigaon are using the root of the Bilabana tree as an anthelmintic. This is effective for all group of livestock. One part of the root is administered to a calf for three days. They caution for dose alteration. It is an irritant, and should be administered by a bamboo tube inserted into the oesophagus, or the food tube.

The villagers of Naskaipadar, administer the Randoli (*Hesperethusa crenulata*) fruit cuticle with its hair is mixed in jaggery. For a buffalo calf, four fruits form one dose.

This is effective for roundworms in buffalo calves. It is an irritant, and should be administered through a bamboo tube only.

Plate 155: Administrating herbal anthelmintics to claves through bamboo tube, Ranagiri

Two parts of the Bhoielimbo (*Androgphis paniculata*) root, one part of the Kambli Kamri (*Eclipta alba*) root and one part of Katagira (*Toona cilliata*) root are mixed. They are very effective as anthelmintics. The Katagira alone is also effective.

Against Lice

A paste is made out of the roots of the Konjapati (*Euonymus glaber*) tree and kept in a pot overnight. If applied on the lice-infested skin before a bath, it eliminates lice.

Otherwise, Bhoie limba leaves are mixed with *Munda Turmeric* (at germination stage) and applied with oil to lice affected hairs for one day. It can be applied on human body also. It will remove all lice.

The root *Heterofragma roxburgii*, locally called Vaira by tribals of Semiliguda, is effective in eliminating lice and ticks on animal body (Behura, 2002, personal communication).

On Wounds

- Kalagacha (*Gouania leptostacheya*), when made to paste, forms a gum like substance, which can be applied over infected wounds. It will heal wounds quickly.

- A paste made from the roots of lau kauti (*Disporum calcaratum*) can be applied to wounds for early healing.
- Daklikanda (*Ghusri oakli*) is a creeper with anti-bacterial property. It is generally given for 1 to 3 days.

Fracture

If there is a fracture on the right side, the juice of a particular plant is put into the left ear, and in the right ear for a fracture on the left side. Medicine is applied over the fractured area and it is plastered with bamboo flints. This is the universal treatment for fracture among Koraput tribes and they claim a very high rage of success. This needs to be investigated.

Maggot Wounds

Maggot wounds can be treated by the following:

- Paste made from Kanchakardi tubers (*Colocasia esculenta*) to be applied on maggot wounds. These tubers look like yam tubers.
- Bonovalia (*Semecarpus anacardium*) seed oil should be to applied on maggot wounds.
- When seven pieces of straw are mixed in three kulia (*Litsea poyantha*), a creeper, or seven grains of common salt are mixed in Kulia, and administered orally, maggots are known to die.
- Mayur mundi (*Elephantopus scaber*) is another plant whose roots if given orally, kills maggot.

Kohler-Rollefson and Braunig (1998) have described the need and approach to apply anthropological veterinary medicine, particularly among the pastoral people of Rajasthan. Cross-cultural ethnobotanical studies of north India to identify common plants among different tribal communities was reported by Jain and Saklani (1992); Saklani and Jain (1994 and 1996).

Later a comparative ethnobotany of India and Latin America forum was established. It identified 650 plants common to both (Chopra et al., 1956, Jain 1991, Singh et al., 1983). This is an upcoming branch of alternative veterinary medicine to be applied by community livestock link workers, which will be the trend in the future.

Further research is needed to establish local and botanical names to see whether these have been covered among the Indian medicinal plants. Then some progress can be said to have been made. However, ethno-medicine has an unique role, whether used alone or with new world medicines.

Threshold Levels of Livestock and Poultry in Tribal Farming

The 'threshold level' in livestock and poultry in the tribal farming system is defined as the optimum number of reproducing adult female animals per family. Thus, if the limiting factors in the given environment are altered, the number of animals per family will also change.

The 'limiting factors' in the tribal livestock systems are

- Epidemic diseases, endemic diseases, production diseases, etc.
- Predators
- Thefts
- Limitation of feed, resources (carrying capacity under scavenging conditions)
- Backyard, horticulture, farming interest and season
- Tribal farming system
- Family resource: Family (men and women), time and occupational interest
- Value (utility) need, religious needs, price, market demand
- Emergency sales
- Feuds with neighbours/villagers

The limiting factors that are associated with high losses are the 'primary limiting factors'. They dictate the unit level

It can also be said that the 'threshold level' is the stage where 'loss absorption is ensured' in the family. However, once the loss inflicting limiting factors are addressed, the threshold level is enhanced to a stage of higher number of reproducing female units and production. At this stage, the feed resource is the 'most limiting factor'. And this, therefore, determines the 'threshold level'.

Prolificacy, Species and Threshold Level

Prolificacy is the key factor in tribal livestock farming. Prolificacy in a species, coupled with scavenging efficiency is most desirable. The livestock species and breeds which attain early maturity, and have short gestation periods, a high number of progeny at a time, with short inter-birth period, are classified as prolific. The indigenous poultry followed by local pigs, goats and sheep genotypes come under this category. The threshold level is increased and that leads to an efficient system. It also ensures highest loss absorption ability, leading to successful as well as sustainable farming system. Prolific sheep breeds in coastal Orissa like Kuzi, Erka or Chillika and goat breed like Black Bengal have special ability to breed fast and regain the population size soon after serious onslaugt due to flood and cyclone.

Tribal Family Livestock Species

POULTRY

With limited backyard space, poultry numbers depend on feeding resources.

The scavenging food resource is a bio-production unit. It is dependent on the total area, the quantity and quality of manure from livestock and household wastes. The season, moisture as well as shade have a bearing over type of insects and fly population that are harbouring in the manure pits. The hidden unicellular component is a vital factor in the production system.

The threshold level for poultry is that tribal families often keep a two-hen unit.

The optimum conditions for poultry rearing are during September to March, when the weather is good and plenty of bio-mass is available after the rains.

Under good conditions, each hen lays about 10 to 15 eggs per clutch, and about 3 clutches per annum. The general mortality of chicks leaves only 6 to 8 chicks alive at weaning time. The optimum survivability is realized from September to March, only when 80-90 per cent hatching rate is observed.

If an efficient prophylactic vaccination programme is maintained, the number of surviving chick will rise dramatically and the family will be encouraged to keep more hens, e.g. a four-hen unit. If more families in a village are keeping a larger number of chicks, feed supplementation is necessary. This situation is termed semi-scavenging.

PIGS

Pigs in tribal villages are scavengers within the limits of the village boundary and almost all the pigs are of the local type. Especially in hill regions both Scheduled Tribes and Scheduled Castes are keeping pigs. The threshold is 'one sow per family'. Local pigs generally mature at about six months of age and a sow gives birth to 5 to 12 piglets. Normally, two litters are produced per year.

The single most limiting factor is losses due to viral swine fever with a mortality rate of up to 90 per cent. Lack of feed resources is another limiting factor.

Nevertheless, since local pigs are not competing with humans for food, they certainly have a place in such a farming system. And, if the feed supplement can be provided, there is scope to improve production.

SHEEP AND GOAT

The feed resource of sheep and goats in tribal villages are communal land, forests, fallow land and the sides of the roads.

Sheep depend on grazing land which may limit the production. The goat is dependent on the forest but sooner or later goats depend on various thorny bushes available in waste lands and road sides. Forests and roadside bushes provide good feed to the goats.

Backyard agriculture and horticultural growth may prevent people from keeping goats in villages.

The housing space, security from theft and increased attacks of predators due to loss of forests are limiting factors to threshold levels in hill and forest regions. It may be concluded that goats have a strong environment limiting factor operational in a tribal farming system. The tribals are very sensitive to environmental issues.

Common Plantation Crops Grown in Koraput District, Orissa

English Name	Botanical Name	Oriya & Local Name
Ash gourd*	*Benincasa cerifera*	Pani kakharu
Broad bean****	*Vicia faba*	Simba, Jhata
Banana*	*Musa paradisiaca*	Kadali
Barnyard millet	*Echinochloa colona*	Khira, Suana
Bullock's heart*	*Annona reticulata*	Ram fala
Black mustard	*Brassica nigra*	Tila
Bitter gourd**	*Momordica charantio*	Kalara
Bottle gourd**	*Lagenaria leucantha*	Lau
Brinjal**	*Solanum melongena*	Baigan
Black gram***	*Phaseolus mungo*	Biri
Cauliflower*	*Brassica oleracea*	Fula kobi
Cucumber*	*Cucumis sativus*	Kakudi, Kakdi
Cashew***	*Anacardium occidentale*	Bhalia
Castor**	*Ricinus communis*	Jada
Custard apple*	*Annona squamosa*	Sita phala
Chilli*	*Capsicum frutescens*	Lanka, Maricha
Coconut	*Cocos nucifera*	Nadia
Double bean**	*Phaseolus lunatus*	Lamba simba, Jhata
Drumstick**	*Moringa pterigosperma*	Sajana, Muninga
Date palm	*Phoenix dactylifera*	Khajuri
Elephant ear*	*Colocasia antiquorum*	Saru
Elephant foot	*Amorphophallus campanulatus*	Olua
Finger millet, Ragi****	*Eleusine coracana*	Mandia
French bean*	*Phaseolus vulgaris*	Farasi Simba
Fenugreek	*Trigonella foenumgraecum*	Methi

Great millet***	*Andropogon sorghum*	
Gram, Bengal gram	*Cicer arietinum*	Chana,
Guava*	*Psidium guajava*	Pijuli, Jambo Nakhchana
Garlic*	*Allium satirum*	Rasun
Ginger*	*Zingiber officinale*	Ada
Green gram	*Phaseolus aureus*	Moong
Horse gram***	*Dolichos oiflorus*	Kolatha
Indian mustard	*Brassica juncea Coss*	Rai, sorisa
Italian fox millet	*Setaria italica Beauv*	Kangu, Kangam
Jack fruit****	*Artocarpus integrifolia*	Panasa
Jujube	*Zizyphus jujuba*	Bara koli
Jamun tree	*Syzygium cumyni*	Jambokuli
Little gourd****	*Coccinia indica*	Tundri
Little millet****	*Panicum miliare*	Suana
Lęntil***	*Lens esculenta*	Harad, Kandula
Linseed	*Linum usitatissinum*	Peshi
Lemon	*Citrus limonia*	Lembu
Maize****	*Zea mays*	Macca
Mango****	*Mangifera indica*	Ambo
Mahua tree	*Madhuca indica* or *M. latifolia*	Maula, Mahua
Mulberry	*Morus alba*	Tutu koli
Niger***	*Abtysinnia guzotica*	Alsi
Onion*	*Allium cepa*	Peas, Ulli, Peaja
Potato*	*Solanum tuberosum*	Bilati Alu, Alu
Pumpkin****	*Cucurbita peop*	Bilati Kakharu,
Papaya*	*Carica papaya*	Ambrut bhanda
Pineapple*	*Ananas sativa Schutt*	Sapuri, Saphrd, Sapuri, panasa Kumuda
Raddish*	*Raphanus sativus*	Mula
Ridge gourd*	*Luffa acutangula*	Janhi
Rice, Paddy****	*Oryza sativa*	Dhano
Sweet Potato**	*Ipomoea batatas*	Kandamula
Sorghum***	*Brot*	Juara, Jona

Safflower*	*Carthamus tinctorius*	Kusum
Sesame* Sesamum	*Sesamum indicum*	Rasi
Sisal hemp**	*Agave spp.*	Murga
Sugar cane	*Saccharum officinarum*	Akhu, Sugar cane Danda
Sago palm*	*Caryota urens*	Solopho
Tomato**	*Lycopersicum esculentum*	Bilati baigan, Bhejiri
Turmeric***	*Curcuma longa* *Curcuma domestica*	Haldi
Tamarind****	*Tamarindus indica*	Tentuli
Yam*	*Dioscorea bulbifera* *Dioscorea alata*	Khamba alu
Yam	*Amaranthus tricolor*	Lal khada, Khoda sag
-	*Amaranthus blitum*	Kosala sag
-	*Amaranthus blitum*	Leutia sag

Note: *, **, *** and **** in increasing intensity of crops grown in the area by tribals, depending upon land availability and usefulness.

Bibliography

Alteri, M.A. and M.K. Anderson, 'An ecological basic for the development of alternative agricultural systems for small farmers in the Third World', *American Journal of Alternative Agriculture* 1:1, 1986, pp. 30-5.

Banik, R.L., 'The edibility of shoots of Bangladesh bamboos and their continuous harvesting effect on productivity', *Bangladesh Journal of Forest Science* 26:1, 1997, pp. 1-10.

Behura, N.K., 'Shifting Cultivation—A cause of environmental degradation', *Adibasi*, vol. 30, no. 4, 1990, pp. 1-7.

———, 'Tribes and the Forest—An Overview', *Adibasi*, vol. 32, no. 1, 1992, pp. 15-19.

Bell, R.C.S., *Orissa District Gazetteer (Koraput)*, Cuttack Government Press, 1945.

Benerjee, Sukumar, *Ethinographic study of the Kuvi Kondha*, Culcutta: Anthropology Survey of India, 1969.

Bhati, J.P. and D.V. Singh, 'Women's contribution to agricultural economy in hill regions of northwest India', *Economic and Political Weekly*, vol. 22, no. 17, 1987, pp. 7-11.

Bhattacharya, S., *Ollari—A Dravidian Speech,* Memoir no. 3, Calcutta: Anthropology Survey of India, 1957.

Biggelaar, Den. C., 'Farming systems development: Synthesizing indigenous and scientific knowledge systems', *Agriculture and Human Values*, vol. 8, nos. 1-2, 1991, pp. 25-36.

Boal, M. Barbara, *The Kondhs: Human Sacrifice and Religious Change, Bhubaneswar: Modern Book Depot, 1982.*

Brown, Lester, *Who Will Feed China*?, New York: W.W. Norte and Co., 1995.

Cain, Rev. J., 'The Koi, a Southern Tribe of the Gond', *Journal of the Royal Asiatic Society, Bengal*, vol. 13, 1881, pp. 410f.

Campbell, John, 'A personal narrative of thirteen years' service among the wild tribes of Kondistan, for the suppression of Human Sacrifice', London, 1864.

Caroll, C.R. and M.A. Altieri, *Why study traditional agriculture? Agroecology*, New York: McGraw Hill Inc., 1990.

Cashman, K., 'Systems of knowledge as systems of domination: the limitations of established meaning', *Agriculture and Human Values*, vol. 8, nos. 1-2, 1991, pp. 49-58.

Census 2001 Report, Orissa State, Bhubaneswar.

Chauhan, S.K., D.S. Thakur and T.V. Moorti, 'Farm income and consumption disparity on tribal Gaddi Farms in Himachal Pradesh', Indian Journal of Animal Science, vol. 61, no. 11, 1991, pp. 1226-9.

Chopra, N.R., S.L. Nayar and I.C. Chopra, *Glossary of Indian Medicinal Plants*, New Delhi: CSIR, 1956.

Das, G.N., 'A Note on the Hill Bondo of Koraput District (Orissa)', *Banyajati*, vol. 2, 1956.

Dasgupta, C.C., 'Human Sacrifices and the Sabaras of Orissa', *Man*, vol. 38, no. 15, 1938.

Das Patnaik, P.S. and B.B. Mohanty, 'Dongria Kondh of Orissa—Then & Now', *Adibasi*, vol. 30, no. 2, 1990, pp. 28-41.

De Bach, P., *Biological Control of Insect Pests & Weeds*, London: Chapman & Weed, 1964.

Dhebar Commission Report, New Delhi: Govt. of India, 1960-1.

Dixit, L.N., Personal Communication on *Cascuta* Control, IKS, 2001.

Elwin, Verrier, *The Muria and their Ghotul*, Delhi: Oxford University Press, 1947.

———, *Bondo Highlanders*, Bombay: Oxford University Press, 1950.

———, *Specimens of the Oral Literature of Middle India* (*Tribal Myths of Orissa*), London: Oxford University Press, 1954.

———, *The Religion of an Indian Tribe*, Bombay: Oxford University Press, 1955.

Fernandes, W., G. Menon and P. Viegas, *Forests, Environment and Tribal Economy: Deforestation, Impoverishment and Marginalisation in Orissa*, New Delhi: Indian Social Institute, 1988.

Francis, W., *Gazetteer of the Vizagapatam District*, vol. I, Madras, 1907.

Furer-Haimendorf, C. Von, 'Megalithic Ritual among the Gadabas and Bondos of Orissa', *Journal of Asiatic Society of Bengal*, vol. 9, 1943, pp. 149-78.

Gangwar, A.K. and P.S. Ramakrishnan, 'Agriculture and Animal Husbandry among the Sulungs and Nishis of Arunachal Pradesh', *Social Action*, vol. 37, no. 4, 1987.

———, 'Cultivation and use of lesser-known plants of food value by tribals in Northeast India', *Agriculture Ecosystems and Environment*, vol. 25, nos. 2-3, 1989, pp. 253-61.

Govindaswamy, S. and A. Krishnamurthy, 'The occurrence of Japonica type of rice grains in Jeypore tract', *Rice News Teller*, vol. 6, no. 4, 1958, pp. 22-29.

Govindaswamy, S., A. Krishnamurthy and N.S. Shastry 'The role of introgression with the varietal variability in rice in Jeypore tract of Orissa', *Oriza*, vol. 3, no. 1, 1966, pp. 74-85.

Grigson, W.V., *The Maria Gonds of Bastar*, Delhi: Oxford University Press, 1938.

Gueye, E.F., 'Poultry plays an important role in Africa', *World's Poultry Science Journal*, vol. 54, no. 1, 1989a, pp. 73-86.

Gupta, D.D. and P.K. Pal, 'Indigenous knowledge base and new agricultural research paradigm', *Journal of Interacademica*, vol. 1, no. 4, 1997, pp. 363-9.

Hahn, N., 'Backyard gardening: a food security system managed by women', *Entwicklung-Landlicher-Raum*, vol. 25, no. 1, 1991, pp. 24-7.

Hahn, N.D., 'Compound and Household Farming: A Sustainable System for African Agriculture', in *Sustainable Agricultural Systems*', ed. Edwards, Lal, Madden, Miller and House, Ankey, Iowa: Publication Soil and Water Conservation Society, 1990, pp. 504-13.

Harlan, J.R., 'Distribution and Utilisation of Natural Variability in Cultivate Plant', *Brookhaven Symposium Biological*, vol. 9, 1956, pp. 191-208.

Harsh, N.S.K., C.K. Tiwari and B.K. Rai, 'Forest fungi in the aid of tribal women of Madhya Pradesh', *Sustainable Forestry*, vol. 1, no. 1, 1996, pp. 1-15.

Hayavardana Rao, C., 'The Gonds of Eastern Ghats', in E. Thurston, *Caste and Tribes of Southern India*, Madras, vols. 1-7, 1907-9.

Henderson, C.A., quoted by J.C. Monoly, *Census of India*, vol. 12, no. 1, 1911, p. 164.

Jain, S.K. and A. Saklani, *Cross-cultural ethnobotanical studies in the northeast India*, Delhi: Deep Publications, 1992.

Jain, S.K., 'Wild Plant Foods of the Tribals of Bastar, M.P.', *Proceedings of National Institute for Scientific India*, 1963-4.

———, *Dictionary of Indian Folk Medicine and Ethnobotany*, Delhi: Deep Publications, 1991.

Jain, Vinita, 'Nature and extent of involvement of men and women in animal husbandry operations', *Indian Dairyman*, vol. 44, no. 7, 1992, pp. 332-7.

Khoshoo, T.N., *Conservation of Biological Diversity Science and Politics, in 40 Years of Research—A CSIR Overview*, ed. N.R. Rajagopal, New Delhi: CSIR, 1987, pp. 347-65.

Kitalyi, A.J., 'Village Chicken Production Systems in developing countries: What does the future hold', *World Animal Review*, vol. 2, 1997, pp. 48-53.

———, 'Family poultry management systems in Africa', The First INFPD/FAO', 1998 (www.fao.org/ag/aga/agap/gpa/fampol/leadpap3.htm)

Kohler Rollefson, Ilse and Braunig Juliane, 'Anthropological Veterinary Medicine: The need for indigenizing the curriculum', 9th Conf., *AITVM Conference in Harare*, 14-18 September 1998.

Kornel, Das, *Tribal Animal Husbandry Systems and Animal Genetic Resources in Southern Orissa*, Monogram 2, Sunabeda, Orissa, 1988.

———, *Livestock and Poultry Dynamics in Tribal life—Koraput*, Bhubaneswar: Modern Book Depot, 1999a.

———, *Livestock and Poultry Genetic Resources in Orissa*, vol. I, Bhubaneswar: Modern Book Depot, 1999b.

———, *Tribal Cultural Heritage and Cult—Gutob Gadaba Tribe of Orissa*, Bhubaneswar: Modern Book Depot, 1999c.

———, *The Koya tribe and Bangladesh Refugees in Dandakaranya* (forthcoming).

Kuit, H.G., A. Tragore and R.T. Wilson, 'Livestock Production in Central Mali: Ownership, Management and Productivity of Poultry in the Traditional Sector', *Tropical Animal Health and Production*, vol. 18, 1996, pp. 222-31.

Lenka, D., *Agriculture in Orissa*, Delhi: Kalyani Publishers, 2001.

MacArthur, J.D., 'Some characteristics of farming in a tropical environment', quoted in Hans Rutherberg (ed.), *Farming System in the Tropics*, 5th edn., Oxford: Clarendon Press, 1980, p. 19.

MacPherson, S. Charles, 'An Account of the Religion of the Kondhs in Orissa, Report by Agent for Suppression of Meriah Sacrifice and Female Infanticide in the Hill Tracts of Orissa', London: Royal Asiatic Society, 1841, pp. 216-74.

Maheshwari, J.K., 'Ethnobotanical documentation of primitive tribes of Madhya Pradesh', *Abst. 4th International Congress of Ethnobiology*, Lucknow, 1994.

Maikhuri, R.K., 'Eco-energetic analysis of animal husbandry in traditional societies of India', *Energy*, vol. 17, no. 1V, 1992, pp. 859-67.

Maikhuri, R.K. and A.K. Gangwar, 'Ethnobiological notes on the Khasis and Garo tribes of Meghalaya, Northeast India', *Economic Botany*, vol. 47, no. 4, 1993, pp. 345-57.

May, A., 'Notes on the Bondas of Jaypur', *The Indian Antiquary*, vol. II, 1873, p. 236.

Menon, G.S., 'Tribal women: Victims of the development process', *Social Action*, vol. 37, no. 4, 1987, pp. 373-90.

Miracle, M.P., *Traditional agricultural methods in the Congo Basin* (mimeo), Standford University, Ford Research Institute, 1964.

Mishra, B.K. and P.S. Ramakrishnan, 'Energy flow through a village ecosystem with slash and burn agriculture in north-eastern India', *Agricultural System*, vol. 9, 1982, pp. 57-72.

Mishra, R.C., Personal communication on rice beans, 2000.

Mishra, K., *Women in Tribal Community: a study of Arunachal Pradesh*, Delhi: Vikas Publishing House, 1991.

Mohanty, B.B., 'Techno-Economic Dimensions of a Tribal Village in Kalahandi District', *Orissa, Adibasi*, vol. 31, nos. 3 and 4, 1991, pp. 31-8.

Mohanty, Gopinath, *Poraja*, tr. B.K. Das, Delhi: Oxford India, 1997.

Mohanty, S.C., 'Paroja', in *Tribes of Orissa*, Bhubaneswar: Harijan, Tribal & Welfare Department, Govt. of Orissa, 1990, pp. 221-9.

Mohapatra, C.R., 'Podu: An Ecological Hazard', *Adibasi*, vol. 32, no. 1, 1992, pp. 20-3.

Mohapatro, M., 'A study of the Socio-political organization of the SA-A-RA', *Adibasi*, vol. 7, no. 1, 1965, pp. 57-9.

Nanda, B.N., *Contours of Continuity and Change—The Story of the Bondo Highlanders*, Delhi: Sage Publications, 1994.

Nayak, P.K., *Blood, Women and Territory: An Analysis of Clan Feuds of Dongria Kondhs*, Delhi: Reliance Publishing House, 1989.

Nayar, N.M., *Origin and Cytogenetics of Rice: Advances in Genetics*, vol. 17, New York, 1973.

Niggemeyar, Hermann, *Kuttia Kond: Dschungel-Bauern in Orissa*, Johann Wolfgang Goethe Universitat, Frankfurt-am-Main, 1964.

Oberoi, R.C, T.V. Moorti and A.K. Sharma, 'Comparative economics of sheep and goat rearing (A study of tribal farm in Western Himalayas)', *Livestock Advisor*, vol. 17, no. 1, pp. 25-30.

Oka, H.I. and W.T. Chang, 'Rice varieties intermediate between wild and cultivated forms and the origin of japonica type', *Botanical Bulletin*, New York: Acadimic Press, vol. 3, 1962, pp. 109-31.

Osunda, M.A.A., 'Indigenous climate knowledge and agricultural practice in south-western Nigeria', *Malaysian Journal of Tropical Geography*, vol. 25, no. 1, 1994, pp. 1-28.

OUAT Report, *Production constraints and research prioritization (Orissa)*, Bhubaneswar: Directorate of Research, Orissa University of Agricultural Technology, 1996.

Panda, J.K., 'Changing Saora culture', *Adibasi*, vol. 7, no. 1, 1965, pp.17f.

Panda, Niranjan and Gurdev S. Khush, *Host Plant Resistance to Insects*, U.K and Phillipines: CAB International and International Rice Research Institute, 1995.

Pandey, A.K., *Shifting cultivation in Northeast India: problems and possible remedies*, Jorhat: Institute of Rain & Moist Deciduous Forest Research, 1996.

Parasar, B.B., Personal communication, 1992.

Pastakia, A.R. and N. Hart, 'SRISTI initiatives for sustainable agriculture and rural development: a response to post-Cold War challenges in India', *Rural Development International Workshop*, Godollo, Hungary, 1996, pp. 117-27.

Patnaik, N. and P.S. Das Patnaik, *The Kondh of Orissa*, Bhubaneswar: Tribal & Harijan Research-cum-Training Institute, 1982.

Patnaik, N., 'Shifting Cultivation in Orissa', *Adibasi*, vol. 16, no. 4, 1977.

———, 'Tribes of Koraput district (their pre-agricultural economic life)', *Adibasi*, vol. 18, no. 1, 1988, pp. 1-18.

———, B. Chowdhury and P.S. Das Patnaik, *The Bondos and their Response to Development*, Bhubaneswar: Tribal & Harijan Research cum Training Institute, 1984.

Patnaik, U.N., 'The Saora of Ganjam Hills', *Adibasi*, vol. 5, no. 1, 1963, pp. 6-13.

———, 'Position of Women in Kondh Society', *Adibasi*, vol. 26, no. 4, 1989, pp. 40-1.

Prasad, Ram, *Mahuwa the Tree of Poor*, Dehradun: International Book Distributors, 1993.

Rajasekaran, B., D.M. Warren and S.C. Babu 'Indigenous Natural Resource Management Systems for Sustainable Agricultural Development—A Global Perspective', *Journal of Inter-national Development*, vol. 3, no. 4, 1991, pp. 387-402.

Ramadas, G., 'The Gadabas', *Man in India*, vol. 11, 1931, pp. 160-73.

———, 'Porajas III', *Man in India*, vol. 16, 1936, pp. 243-58.

Ramakrishanan, P.S. and A.K. Gangwar, 'Ecosystem function in a Khasi village of the desertified Cherrapunji area in north-east India', *Proceedings of the Indian Academy Plant Science*, 99, vol. 3, 1989, pp. 199-210.

Ramakrishnan, P.S. and R.K. Maikhuri, 'Ecology and management of Jhum cultivation in the Himalayas', 3rd International Dry Land Congress, New Delhi, 1988, pp. 26-37.

———, 'Ecological analysis of a cluster villages emphasizing land use of different tribes in Meghalaya in north-east India', *Agric. Ecosys. Environ*, vol. 31, 1990, pp. 17-27.

Ramamurti, G.V., *A Manual of the Sora (Savara) Language*, Madras, 1931.

Ramiah, K., *Rice in Madras*, Madras: Govt. Press, 1937.

———, 'Rice Breeding and Genetics', *Scientific Monograph*, no. 19, New Delhi: Indian Council of Agricultural Research, 1953.

Ramiah, K. and R.L.M. Ghose, 'Origin and Distribution of Cultivated Plants of South Asia Rice', *Indian Journal of Genetic and Plant Breeding*, vol. 11, 1951, pp. 7-13.

Ramlah, A.H. and H. Kassim, 'Production performance of village chicken', in *Proceedings of the National IRPA Seminar*, vol. 2, 1992, pp. 496-7.

Randhawa, M.S., *A History of Agriculture in India*, vol. 1, New Delhi: Indian Council of Agricultural Research, quoted Watt, 1980.

Rathore, O.S. and K.S.K. Gaur, 'A comparative study on contribution and time utilization pattern of farm women in tribal and non-tribal areas', *Mysore Journal of Agricultural Sciences*, vol. 30, no. 1, 1996, pp. 87-92.

Routray, S., 'The Status of Women among the Dongria Kondh', *Adibasi*, vol. 27, nos. 2 and 3, 1987, pp. 31-7.

Roy, J.K. and P.K. Rao, 'Investigation on the diet of the Muria of Bastar District', *Bulletin*, Department of Anthropology, Govt. of India, vol. 6, 1957, pp. 33-45.

Roy, K.C. and C.A. Tisdel, 'Technological change, environment and poor women, especially tribal women, in India', *Savings & Development*, vol. 17, no. 4, 1993, pp. 423-39.

Roy, S.N., 'The Savaras of Oriss', *Man in India*, vol. 7, 1927, p. 287.

Rutherberg, Hans, *Farming Systems in the Tropics*, 5th edn., Oxford: Clarendon Press, 1980.

Sahu, A.C., 'Shifting cultivation to Horticulture—A case study of the Dongria Kondhs, Development Agency (Kurli)', *Adibasi*, vol. 32, no. 1, 1992, pp. 30-41.

Sahu, Dibakar, 'Rainfed farming in Tribal Areas for Environmental and Food Security', *Adibasi*, vol. 32, no. 1, 1992, pp. 24-9.

Saklani, A. and S.K. Jain, *Cross-cultural Ethnobotany of North-east India*, Delhi: Deep Publications, 1994.

———, 'Credibility of folk claims of north-eastern Himalaya and north-western India', in S.K. Jain, ed., *Ethnobiology in Human Welfare*, Delhi: Deep Publications, 1996, pp. 136-9.

Salem, H.B. and M.B. Hammouda, 'Goat rearing in the humid and sub-humid region of Bizerte', *Medit.*, vol. 6, no. 2, 1995, p. 59-61.

Samantray, B.M., 'Personal communication on Nodia rice', 1999.

Samantray, H.M., 'Personal communication on Patangi rajama', 1992.

Santapu, H., *India: The Land and the People - Common Trees*, New Delhi: National Book Trust of India, 1966.

Sarma, N.N. et al., 'Improved practice in place of shifting cultivation and its effect on soil properties at Diphu in Assam', *Indian Journal of Agricultural Science*, vol. 65, no. 3, 1995, pp. 196-201.

Saxena, H.O. and M. Bralman, *The Flora of Orissa*, Bhubaneswar, RRL & OFDC Ltd., 1994.

Schafer, J., 'Utilizing Indigenous Agricultural Knowledge in the Planning of Agricultural Research Projects Designed to Aid Small-scale Farmers', in D.M. Warren, L.J. Slikkerveer and S.O. Titilola, eds., *Indigenous Knowledge Systems: Implications for Agriculture and*

International Development, Studies in Technology and Social Change, no. 11, Ames, Iowa: Technology and Social Change Program, Iowa State University, 1986, pp. 116-20.

Sere, C. and H. Steinfeld, 'World Livestock Production Systems: Current Status, Issues and Trends', *Animal Production and Health Paper No. 127*, Rome: FAO, 1996.

Sharma, B.D., 'Forests, tribal economy and regional development', *Bulletin of the Tribal Research and Development Institute,* Bhopal, vol. 13, 1985, pp. 1-19.

Sharma, S.D., 'Resource Centre for Farmers' Rights', in M.S. Swaminathan, ed., *Agro-biodiversity and Farmers' Rights*, Delhi: Konark Publishers, 1996, pp. 177-9.

———, Smita Tripathy and Pratima Gurung, 'Jeypore tract—Case study', in M.S. Swaminathan, ed., *Gender Dimensions in Biodiversity Management*, Delhi: Konark Publishers, 1998.

Singh, Bhupinder, 'The Tribal Problem: Some reflections in steps towards Tribal Development', in Singh Bhupinder, ed., *Training of Trainers in Tribal Development Administration*, Training Division, Department of Personnel & Administrative Reforms, Govt. of India, New Delhi, vol. 1, 1982, pp. 36-51.

Singh, A.K., 'Status of tribal women in India', *Social Change*, vol. 23, no. 4, Delhi, 1993, pp. 3-143.

———, *The Saora Highlander Leadership and Development*, Delhi: Somaiya Publication, 1984.

———, 'Forest policy and tribal development', *Adibasi*, vol. 27, no. 1, 1987, pp. 5-9.

Singh, U., A.M. Wadhwani and B.M. Johri, *Dictionary of Economic Plants in India*, New Delhi: Indian Council of Agricultural Research, 1983.

Sitapati, G.V., 'The Soras and their country', *Journal of Andhra Historical Research Society*, vol. 12, 1938, pp. 57f.

Sonaiya, E.B., 'The context and prospects for development of smallholder rural poultry production in Africa', *Proceedings*: *CTA International Seminar on Smallholder Rural Poultry Production*, Thessaloniki, Greece, vol. 1, 1990, pp. 35-52.

Sonaiya, E.B., R.D.S. Brakaert and E.F. Gueye, 'Research and Development Options for Family Poultry', introductory paper to the first INFPD/FAO Electronic Conference on Family Poultry, 1998, pp. 1-18.

Southwood, T.R.E., 'The dynamics of insect populations', in D. Pimentel, ed., *Insects, Science and Society*, New York: Academic Press, 1975, pp. 151-99.

Spencer, J.E., *Shifting Cultivation in South Eastern Asia*, California: University of California Press, 1966.

Starling, A., 'An Account, Geographical, Statistical and Historical Orissa Proper, or Cuttack', *Asiatic Researchers*, vol. 15, 1825, p. 43.

Stuart, 1891 quoted by E. Thurston and K. Rangachari, *Castes and Tribes of Southern India*, Madras, 1909, pp. 1-7.

Swaminathan, M.S., *Science and Conquest of Hunger*, Delhi: Concept Publishing Co., 1982.

———, *Science and Integrated Rural Development*, Delhi: Concept Publishing Co., 1982.

———, *Agrobiodiversity and Farmers Rights*, Delhi: Konark Publisher, 1996.

———, *Sustainable Agriculture—towards food security*, Delhi: Konark Publishers, 1996.

Taylor, Rev.W., 'Catalogue Raisonns', *Oriental Manuscripts*, 1862.

Tewari, D.N., 'Bamboo as Poverty Alleviator', *Indian Forester*, vol. 114, no. 10, 1988, pp. 610-12.

Thapa, B., Sinclair et al., 'Nepalese farmers knowledge of tree fodder resources: implication for research and development', selected papers from the international symposium on Systems-oriented Research in Agriculture and Rural Development, Montpellier, France, 1996, pp. 125-36.

Thear, K., *Free Range Poultry*, 2nd edn., UK: Farming Press, p. 173.

Thurston, E. and K. Rangachari, *Castes and Tribes of Southern India*, vols. 1-7, Madras, 1909.

Thusu, Kidar Nath and Makan Jha, *Ollar Gadaba of Korapu*, Calcutta: Anthropological Survey of India, 1972.

U.N. Conference on Social Development in Copenhagen in March 1995.

Ulaka, Rama Chandra, 'The social life of the Paraja', *Adibasi*, vol. 16, no. 2, 1976, pp. 1-32.

van Emden, H.F., 'The role of uncultivated land in the biology of crop pests and beneficial insects', *Scientific Horticulture*, vol. 17, 1965, pp. 121-36.

Vanek, E., 'Enhancing environmental resource management in developing nations through improved attitudes towards indigenous knowledge systems: the case of the World Bank', in D.M. Warren et al., eds., *Indigenous Knowledge Systems: Implication for Agriculture and International Development, Studies in Technology and Social Change*, no. 1, Ames, Iowa: Technology and Social Change Programme, Iowa State University, 1989, pp. 162-70.

Vavilov, N.I., 'The Origin Variation, Immunity and Breeding of Cultivated Plants', *Chronic Botanic*, vol. 13, 1951, pp. 1-364.

Warren, D.M., L.J. Slikkerveer and S.O. Titilola, eds., *Indigenous Knowledge Systems: Implications for Agriculture and International Development Studies in Technology and Social Change*, no. 11, Ames,

Iowa: Technology and Social Change Program, Iowa State University, 1989.

Warren, D.M. and K. Cashman, 'Indigenous knowledge for sustainable agriculture and rural development', Institute for Environment and Development, Iowa State Unit USA, 1989, SA10, 15pp.

Warren, D.M., L.J. Slikkerveer, and D. Brokensha, eds., *The Cultural Dimension of Development: Indigenous Knowledge Systems*, vol. 18, London: Intermediate Technology Publications, 1995.

Watts, Neville A., *The Half-clad Tribals of Eastern India*, Delhi: Orient Longman, 1970.

Watherhogg, J., J. Dixon and K. d'Alwis, *Global Farming Systems Study: Challenges and Priorities to 2030, Regional Analysis – South Asia*, Rome: FAO, 2001.

Williams, G.E.S., 'Small holder poultry production in Ghana', in *CTA Seminar Proceedings*, Small holder Rural Poultry Production, 1990, Thessaloniki, Greece, pp. 89-97.

Index